Creep Behaviour in Cracked Sections of Fibre Reinforced Concrete

RILEM BOOKSERIES

Volume 14

RILEM, The International Union of Laboratories and Experts in Construction Materials, Systems and Structures, founded in 1947, is a non-governmental scientific association whose goal is to contribute to progress in the construction sciences, techniques and industries, essentially by means of the communication it fosters between research and practice. RILEM's focus is on construction materials and their use in building and civil engineering structures, covering all phases of the building process from manufacture to use and recycling of materials. More information on RILEM and its previous publications can be found on www.RILEM.net. Indexed by SCOPUS and Springerlink.

rilem

More information about this series at http://www.springer.com/series/8781

Pedro Serna · Aitor Llano-Torre
Sergio H.P. Cavalaro
Editors

Creep Behaviour in Cracked Sections of Fibre Reinforced Concrete

Proceedings of the International RILEM Workshop FRC-CREEP 2016

Springer

Editors
Pedro Serna
ICITECH Institute for Concrete Science and Technology
Universitat Politècnica de València (UPV)
Valencia
Spain

Sergio H.P. Cavalaro
Departamento de Ingeniería Civil y Ambiental
Universidad Politécnica de Cataluña (UPC)
Barcelona
Spain

Aitor Llano-Torre
ICITECH Institute for Concrete Science and Technology
Universitat Politècnica de València (UPV)
Valencia
Spain

ISSN 2211-0844 ISSN 2211-0852 (electronic)
RILEM Bookseries
ISBN 978-94-024-1000-6 ISBN 978-94-024-1001-3 (eBook)
DOI 10.1007/978-94-024-1001-3

Library of Congress Control Number: 2016954542

Printed on acid-free paper

This Springer imprint is published by Springer Nature
The registered company is Springer Science+Business Media B.V.
The registered company address is: Van Godewijckstraat 30, 3311 GX Dordrecht, The Netherlands

Preface

Over the past years, there has been a significant effort in the research, development and application of fibre-reinforced concrete (FRC). This has been promoted by the introduction of FRC in major national and international codes of structural concrete (ACI 318, MC 2010 and the incoming EC2 among others). However, the long-term behaviour of FRC under sustained load is still a controversial topic in both scientific and technical forums.

One of the main issues to consider this phenomenon in codes lies in the absence of validated and widely accepted models to predict it and standardized testing methodologies to quantify creep in FRCs. The knowledge of the creep response on cracked FRC elements is essential to ensure the viability of structural applications of this material and has important consequences on structural durability and safety. In 2014, a new technical committee was created in the frame of RILEM: the RILEM TC 261-CCF, entitled "Creep behaviour in Cracked Sections of Fibre Reinforced Concrete". The TC 261-CCF is in charge of compiling information on this topic, as well as of reaching a consensus about creep test methodologies and the considerations for the analysis of the results. One of the proposed activities of this TC was the organization of a RILEM workshop, with the aim of including in a publication all collected updated researches and results on the topic.

With this objective, the *FRC-CREEP 2016—International RILEM Workshop on Creep Behaviour in Cracked Sections of Fibre Reinforced Concrete* took place in the "Universitat Politècnica de València" (Spain) during 9–10 March 2016. The *FRC-CREEP 2016* was proud to welcome 35 attendees from 16 countries, and 20 lectures were presented and discussed. Most of the active researchers in the field took part in this event, which was a great occasion to share experiences and establish points of interest for future studies.

The technical activity of the workshop was scheduled in four general sessions, with an additional conference of the Honorary Chairman Ravindra Gettu and two round tables in order to promote the discussion on two specific aspects: the "Creep Testing Methodologies and Results interpretation" and "When creep in a cracked section of structural FRC elements is an important issue". This book presents the proceedings of the workshop.

I would like to take advantage of this preface to state my sincere gratitude to the following people:

- Ravindra Gettu, for promoting the event as the Workshop Honorary Chairman;
- The round tables' organizers Giovanni Plizzari and Claudio Mazzotti, for accepting the challenge of generating discussion on essential aspects;
- The authors for their invaluable technical contributions and willingness to share their research;
- The scientific committee for their collaboration and dedication to the revision of technical articles, a critical process to ensure a high technical quality and achieve valuable proceedings;
- The workshop sponsors for the generous support and collaboration;
- The "Universitat Politècnica de València" and the "Civil Engineering School" that provided the facilities and the infrastructure to hold the workshop;
- The contribution of the organizing committee, the support group, and specially of Aitor Llano who made this project a success.

Finally, I would like to thank all participants, whose interest and enthusiasm made this workshop a first-class and enriching event for all.

Thanks to all of you for your contribution.

Valencia, Spain

Pedro Serna
FRC-CREEP 2016 Workshop Chairman

Committees

Organizing Committee

Pedro Serna, Spain (Chairman)
Aitor Llano-Torre, Spain (Secretariat)
Sergio Henrique Pialarissi Cavalaro, Spain

Honorary Chairman

Ravindra Gettu, India

Scientific Committee

Charles Allen, UK
Bryan Barragán, France
Olivier Bayard, France
Stefan Bernard, Australia
Benoît Bissonnette, Canada
William Peter Boshoff, South Africa
Wolfgang Brameshuber, Germany
Nicola Buratti, Italy
Jean-Philippe Charron, Canada
Christian Clergue, France
Simon Cleven, Germany
Frank Dehn, Germany
Clelia Desmettre, Canada
Marco Di Prisco, Italy
Terje Kanstand, Norway
Wolfgang Kusterle, Germany

Sukmin Kwon, Japan
Catherine Larive, France
Ingemar Lofgren, Norway
José R. Martí-Vargas, Spain
Claudio Mazzotti, Italy
Sandro Moro, Italy
Tomoya Nishiwaki, Japan
Benoit Parmentier, Belgium
Giovanni Plizzari, Italy
Steven Pouillon, Belgium
Pierre Rossi, France
Romildo D. Toledo Filho, Brazil
Lucie Vandewalle, Belgium
Emilia Vasanelli, Italy
Gerhard Vitt, Belgium
Rutger Vrijdaghs, Belgium
Sébastien Wolf, Luxemburg
Raúl Zerbino, Argentina

Institutions and Sponsors

Organized By
Universitat Politècnica de València UPV
Institute of Science and Technology of Concrete—ICITECH

Supported By
RILEM
Generalitat Valenciana—Conselleria d' Educació, Investigació, Cultura i Esport

Sponsored By
Saint-Gobain Seva—FIBRABLEX

Contents

Part I
Keynote Lectures

Factors Influencing Creep of Cracked Fibre Reinforced Concrete: What We Think We Know & What We Do Not Know

Ravindra Gettu, Raúl Zerbino and Sujatha Jose

Abstract The significance of the creep behaviour of fibre reinforced concrete (FRC) in the cracked state is a matter of debate. As FRC in service could be in the cracked state, the serviceability and failure will depend on the stability of the cracks, and the alteration of the capacity to transfer stresses. Based on the present knowledge, this paper discusses factors influencing the creep of cracked FRC to facilitate further discussions. Effects of load, temperature and humidity, and the importance of progressive debonding and pull-out, creep of the fibres and crack propagation under sustained loading have been discussed for steel and polymer fibres.

Keywords Fibre reinforced concrete · Creep behaviour · Pull-out · Steel fibres · Polymeric fibres

1 Introduction

Over the past few decades, fibres have been incorporated in concrete in many applications, such as pavements, slabs on grade, tunnel linings and structural elements along with conventional reinforcement, to improve the service life and failure mode of structures. Since the serviceability depends on the stability of cracks and capacity to transfer sustained stresses, the creep behaviour of cracked fibre reinforced concrete (FRC) is significant. The *fib* Model Code 2010 raises concerns about the long-term performance of concrete reinforced with fibres that could be affected by creep.

R. Gettu (✉) · S. Jose
Indian Institute of Technology Madras, Chennai, India
e-mail: gettu@iitm.ac.in

R. Zerbino
CONICET Researcher, LEMIT-CIC, Civil Engineering Department, UNLP, Ensenada, Argentina

P. Serna et al. (eds.), *Creep Behaviour in Cracked Sections of Fibre Reinforced Concrete*, RILEM Bookseries 14,
DOI 10.1007/978-94-024-1001-3_1

As is well-known, creep is defined as the deformation in a solid due to sustained stresses. Creep of plain concrete or the matrix of FRC under compression load occurs at all stress levels and is due to movement of pore water in the paste, compaction of C-S-H sheets and microcracking, especially at relatively higher stresses. Consequently, the main factors that impact the rate of creep in compression are the type of curing, age of loading, temperature and relative humidity of the environment [1]. In plain concrete, the fracture toughness and fracture process zone size (i.e., the region over which energy is dissipated in front of the propagating crack) can decrease as the loading rate decreases. Consequently, creep could decrease the load carrying capacity of a cracked element significantly [2].

Researchers who have studied the long term behaviour of FRC agree on its ability to better sustain the load levels over time, reducing deflection and crack opening compared to concrete without fibres. In the case of cracked FRC, the fibres bridging the cracks are subjected to sustained loading [3], which could lead to progressive creep of the fibre-matrix interface, and debonding and pull-out of the fibre. These phenomena, along with matrix creep and drying shrinkage in the compressive zone ahead of the crack [4, 5], creep in the individual fibre or filament itself and, more importantly, crack propagation [6] can result in crack widening that can affect serviceability, long-term load-carrying capacity and durability [4, 7–11]. The behaviour varies with the fibre type and dosage, as expected [12, 13].

More than 15 research groups in the world are working now on this subject and efforts have been made to arrive at standard creep testing procedures for cracked FRC. At this point of time, when there is considerable information compared to few years ago, it appears relevant to discuss *what we think we know and what we do not know* about the assessment of tensile creep in cracked FRC.

2 Effect of Load, Temperature and Humidity

Studies have shown that the load level (i.e., the ratio of the sustained load applied to the specimen to that required for precracking the specimen) has an important effect on flexural creep response of cracked FRC. It was found that an increase in the load ratio always leads to higher creep strains [14]. Zerbino and Barragán [8] observed that the precrack width and sustained load level are interrelated; a stable creep response is obtained at lower precrack widths, under higher load ratios or at wider cracks, under lower load levels.

Temperature also has a significant influence on the long term behaviour of FRC. The effects of moderate temperatures (20 to 40 °C) and elevated temperatures (about 60 °C) on the long term behaviour have been studied by some researchers [13, 15, 16]. Increase in temperature has an effect on creep performance of steel and polymer fibre reinforced concrete, with higher creep deformation at higher temperature. It was observed that at room temperature and at elevated temperature polymeric fibre reinforced concrete (PFRC) experienced larger creep deformations than steel fibre reinforced concrete (SFRC), especially for larger initial crack widths

(≥ 1.5 mm). However for smaller crack openings, the effect of temperature variation is negligible. The increased creep deformation of steel FRC at moderate to elevated temperatures might be caused by modifications in the long term behaviour of fibre-concrete bond while the creep failure of synthetic FRC might be due to creep deformations of the fibres themselves. Under wetting and drying cycles, the effect of temperature seems to be further aggravated [17]. In general, the ambient humidity also has an effect on the creep of cracked FRC, with deflections increasing with a drier environment, leading to more creep deformation and earlier creep failure.

3 Progressive Pull-Out Mechanisms

Fibre pull-out has been found to be one of the most important factors controlling the creep response of the cracked FRC. The process consists of debonding and frictional pull-out, and depends on the type of fibres, as described schematically in Fig. 1. For hooked-ended steel fibres in cracked concrete [18], the debonding of the interface and fibre pull-out progresses from the crack faces to the ends [19, 20]. The fibre diameter, hook dimensions, inclination to the crack face, interface properties and shear strength are the main factors that influence the energy needed to deform the hook. The pull-out load is highest when the first bend of the fibre is deformed and thereafter the pull-out force decreases, with a slight increase at each deformation of subsequent bends. Obviously, the pull-out resistance is higher for hooked-ended steel fibres than for straight and smooth fibres. Along with the progressive pull-out of the fibres across the crack faces, they also undergo plastic

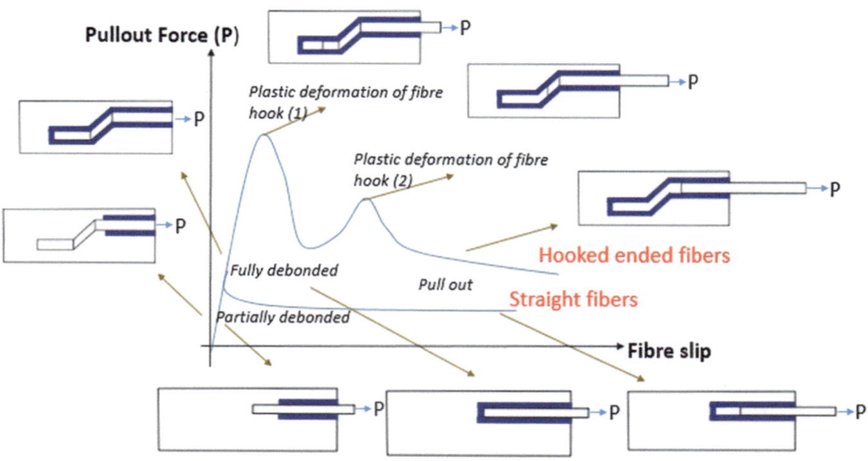

Fig. 1 Typical debonding and fibre pull-outresponse of straight and hooked-ended fibres (adapted from [18])

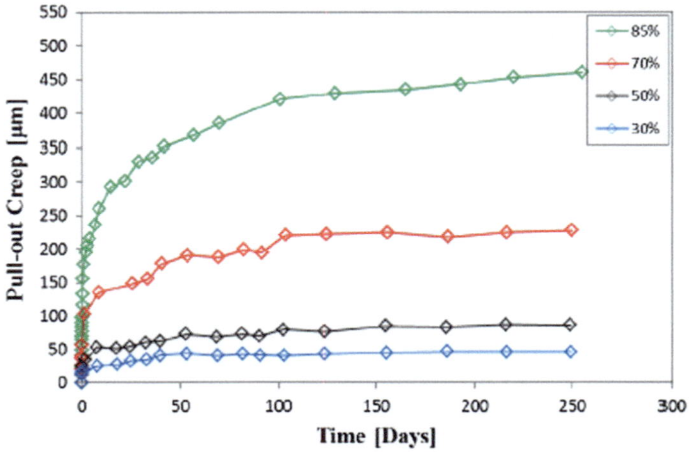

Fig. 2 Single hooked ended fibre response (with permission of Dr. Boshoff from [21])

Fig. 3 Lateral contraction during pull-out

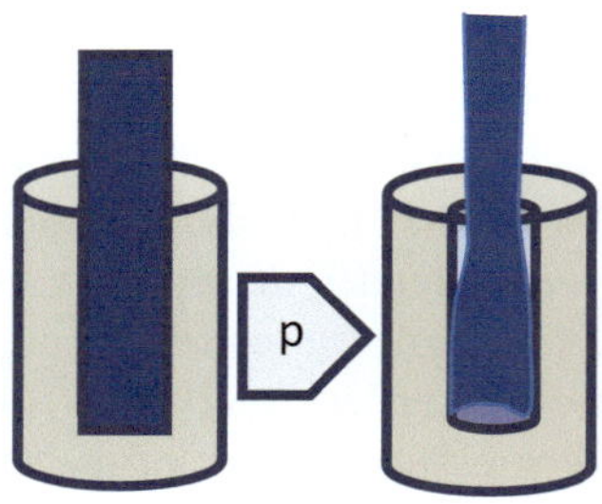

deformation. Nieuwoudt and Boshoff [21] subjected single fibres to sustained load and obtained the responses shown in Fig. 2 for different load levels, with the pull-out displacement clearly increasing with time and load level.

As the crack opens, the fibre is subjected to a tensile stress in the longitudinal direction, which causes Poisson's contraction in the transverse direction, which becomes relevant in the case of polymeric fibres. The Poisson's ratio (ν) is near 0.40–0.45 for polypropylene, which is much greater than in steel (0.27–0.30). As illustrated in Fig. 3, the lateral contraction of the fibre due to the Poisson's effect will facilitate debonding and pull-out. So, fewer thick fibres can be expected to be less effective in resisting long-term pull-out than finer ones (for the same dosage).

3.1 Damage Due to Anchorage Failure in Deformed Fibres

Nieuwoudt and Boshoff [21] recently showed that fracture may occur in the matrix surrounding the fibre as it is pulled-out, especially in the case of deformed steel

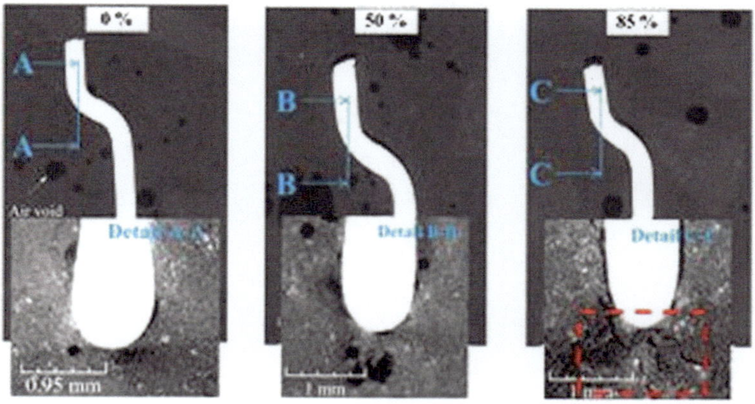

Fig. 4 CT scan images of a single fibre at different stages of pull-out; the lower images are magnifications of the zones indicated in the upper images (with permission of Dr. Boshoff from [21])

fibres (Fig. 4), both hooked-ended, as shown here, or undulated. When such damage progresses with time and load level, due to microcracking and local crushing produced by sustained loading, the pull-out is further facilitated. In the case of straight fibres and lower modulus fibres, like polymeric fibres, such damage is not expected to be significant.

3.2 *Filament or Individual Fibre Creep or Relaxation*

Compared to steel fibres that hardly experience any creep at ambient temperature, filaments of polymeric fibres could experience creep, especially at higher temperatures, due to their viscoelastic nature. Nevertheless, cross-linked and more crystalline polymers are less susceptible to creep. The general trends in creep and relaxation response of polymers are shown in Fig. 5a, b, respectively [22]. When the fibre undergoes creep strain, the crack width will tend to increase. Moreover, in PFRC, the relaxation of the stress in the fibre will decrease the bridging effect, leading to crack widening and propagation. Obviously, such effects will be more prominent as the temperature increases, especially above the glass transition temperature of the polymer.

At this point, it must be emphasized that there are significant differences among polymeric fibres, as a function of the polymer family, shape, surface texture and the production process, and, therefore, generalizations should be made with caution.

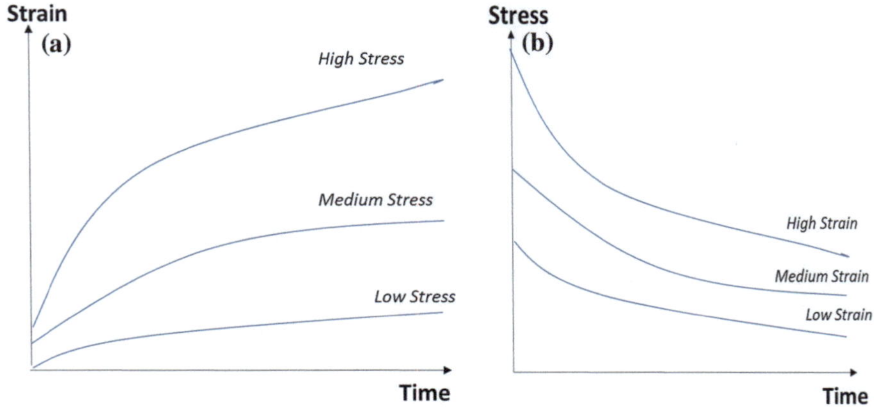

Fig. 5 **a** Creep of a typical polymer under constant stresses; **b** Stress relaxation of a typical polymer under different strain (adapted from [22])

4 Crack Propagation Due to Relaxation of Cohesive Stresses

In plain concrete, the fracture process zone created by rapid loading shrinks due to relaxation of the bridging stresses. This leads to crack propagation that clearly affects load carrying capacity of the concrete (see Fig. 6).

What happens when fibres are present in such a matrix? When a strain-softening type FRC is subjected to constant load, crack propagation and widening with time

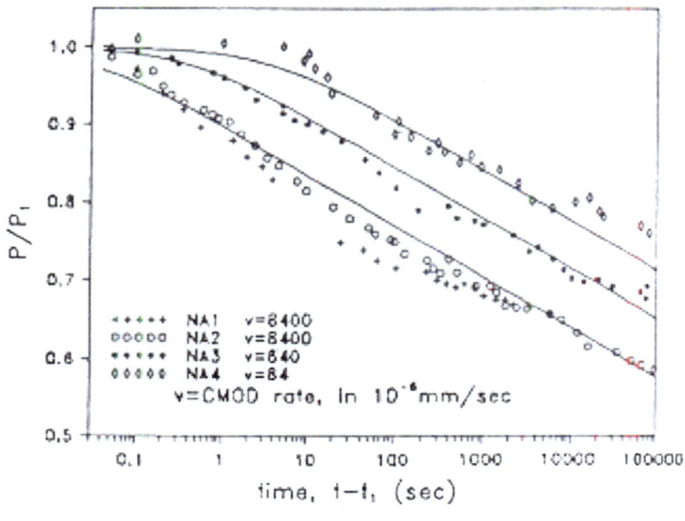

Fig. 6 Relaxation in plain concrete notched beams, in the post-peak regime [2]

can be expected. Eventually, sudden failure may occur after tertiary creep response when the load ratio is high enough. However, when strain-hardening type FRC is subjected to constant load, it is expected that more cracks will widen with time without any failure occurring [23]. This is coherent with the loading level effects discussed in Sect. 2 since the load on the individual fibre, which is bridging the crack, increases as the crack propagates. Figure 7a, b show some responses observed in the RILEM TC CCF Round Robin Test programme for flexural creep for PFRC and SFRC, respectively. It is seen in the plots of crack mouth opening displacement (CMOD) evolution with time, under constant load, that there is a progressive increase in crack opening with some jumps (i.e., sharp increase), each of which indicate sudden crack propagation. In a recent study, Daviau-Desnoyers et al. [24] also concluded that crack propagation is the mechanism that governs the crack widening in FRC due to sustained loading.

In general, when crack propagation leads to failure, the response can be expected to exhibit considerable variability (as in the case of fatigue failure) since the

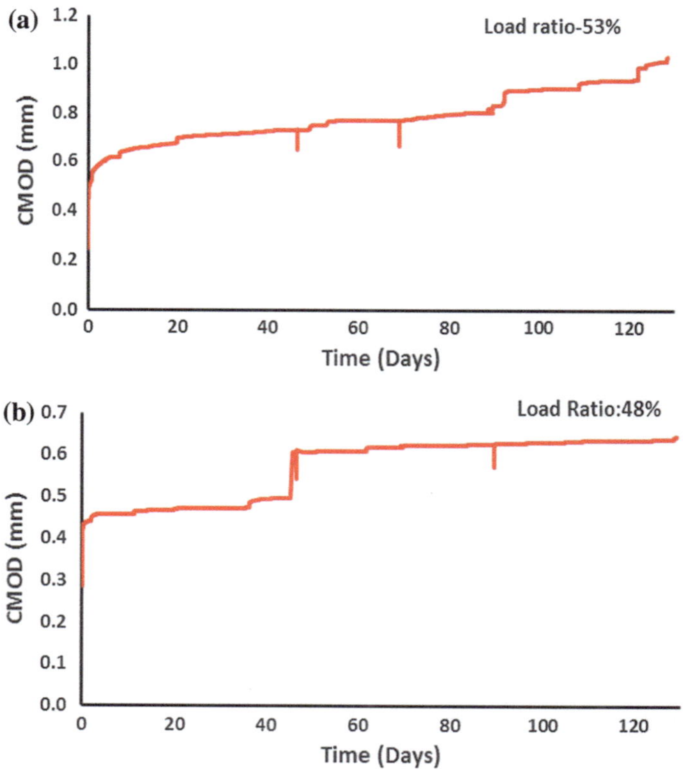

Fig. 7 Responses observed for **a** PFRC and **b** SFRC

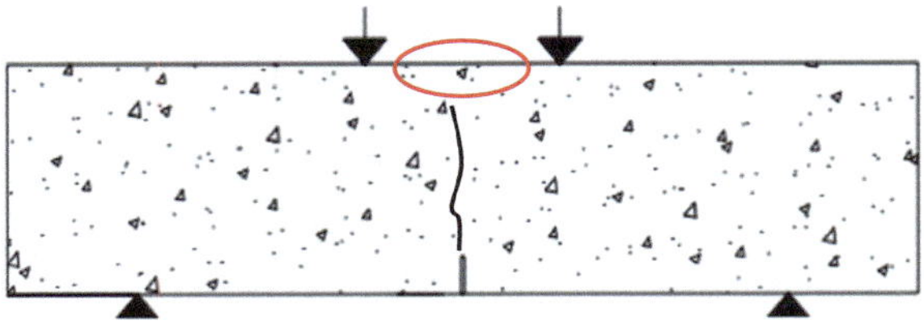

Fig. 8 The compression zone is strongly reduced due to crack propagation

microcracking characteristics, and fibre location and orientation would differ significantly from one specimen to another. However, within the secondary creep regime, the later behaviour in the same specimen may be inferred from earlier time-dependent behaviour [25].

When crack propagation occurs in a typical beam test that is adopted for creep characterization, the compressive stresses increase since the moment of inertia of cracked section (I_{cr}) decreases with time. For example, analysis shows that in a SFRC beam of 150 mm depth, the crack length can be about 100 mm at CMOD = 0.5 mm. Consequently, in this situation, the compression zone would be small and the compressive strains increase as the crack propagates. Therefore, one should be careful while representing the strain increase observed as compression creep (see Fig. 8). Also, the creep coefficient obtained from flexural cracking tests will not have the same relevance as that of linear creep under compression. So formulations homologous to those used until now for compressive creep (of uncracked concrete) would not have any significance in the case of the creep of cracked FRC.

5 Relevance of Wide Cracks

It is clear that crack widening occurs due to relaxation of cohesive stresses and crack propagation. This can be further worsened if ageing affects the bond. At the same time, the increase of the crack width increases the possibility of the tertiary creep to occur. Therefore, the maximum acceptable crack width seems to be more relevant than the maximum increase in crack width that FRC can sustain.

Durability of the steel fibres depends on crack width and is affected more when the fibres are under stress (i.e., stress corrosion); steel fibres may not survive crack widths greater than 0.1 mm. So, it seems to be only of academic interest that we assess behaviour under large crack widths or over many years. Further, one must keep in mind that FRC is generally recommended to improve the concrete durability and if the cracks are very wide, the transport properties are seriously affected resulting in sulphate and chloride attacks. This should be taken into account in the definition of crack width limits for different structural elements.

6 Conclusions

A discussion of the factors affecting the creep behaviour of cracked FRC has been presented. The main conclusions can be listed as follows:

- The creep fracture of FRC is dominated by crack propagation, progressive debonding and pull-out, and fibre creep.
- In general, the creep crack widths increase with loading level, temperature and drying.
- The increase of crack widths due to sustained loading can be expected to be more in PFRC than SFRC.
- The pull-out mechanisms, due to mechanical anchorage and adherence, could have a significant influence on creep behaviour.
- Considering the effects of tertiary creep and durability of the exposed fibres, a maximum acceptable crack width seems to be more relevant in practice than the maximum increase in crack width that the FRC can sustain.

There are, however, many challenges to be addressed in the future that warrant more detailed research.

Acknowledgments Partial funding for this work was provided by the IIT Madras grant CIE/14-15/833/RFER/RAVG on "Study of long-term behaviour of fibre reinforced concrete for tunnel linings and on-grade applications".

References

1. Van Bergen, S.S., Pouillon, S., Vitt, G.: Experiences from 14 years of creep testing of steel and polymer fibre reinforced concrete. In: Serna, P. et al. (eds.) FRC-Creep 2016, Proceedings of International RILEM Workshop on Creep Behaviour in Cracked Sections of Fibre Reinforced Concrete, Valencia, Spain (2016)
2. Bazant, Z.P., Gettu, R.: Rate effects and load relaxation in static fracture of concrete. ACI Mater. J. **89**(5), 456–468 (1992)
3. Vrijdaghs, R., di Prisco, M., Vandewalle, L.: Creep deformations of structural polymeric macrofibres. In: Serna, P. et al. (eds.) FRC-Creep 2016, Proceedings of International RILEM Workshop on Creep Behaviour in Cracked Sections of Fibre Reinforced Concrete, Valencia, Spain (2016)
4. Rossi, P., Charron, J.P., Bastien-Masse, M., Tailhan, J.-L., Le Maou, F., Ramanich, S.: Tensile basic creep versus compressive basic creep at early ages: comparison between normal strength concrete and a very high strength fibre reinforced concrete. Mater. Struct. **47**(10), 1773–1785 (2013)
5. Babafemi, A.J., Boshoff, W.P.: Tensile creep of macro-synthetic fibre reinforced concrete (MSFRC) under uni-axial tensile loading. Cem. Concr. Composites **55**, 62–69 (2015)
6. Kanstad, T., Žirgulis, G.: Long-time creep testing of pre-cracked fibre reinforced concrete beams. In: Proceedings of 8th RILEM International Symposium on Fibre Reinforced Concrete, RILEM PRO88, pp. 195–196. Guimarães, Portugal (2012)

7. Boshoff, W.P., Mechtcherine, V., van Zijl, G.P.A.G.: Characterising the time-dependant behaviour on the single fibre level of SHCC: Part 1: Mechanism of fibre pull-out creep. Cem. Concr. Res. **39**, 779–786 (2009)
8. Zerbino, R., Barragán, B.: Long-term behavior of cracked steel fibre-reinforced concrete beams under sustained loading. ACI Mater. J. **109**(2), 215–224 (2012)
9. Babafemi, A.J., Boshoff, W.P.: Preliminary creep behaviour of polypropylene fibre reinforced concrete (PPFRC) under a high tensile stress, in Advances in Cement and Concrete Technology, Proceedings of International Conference, pp. 887–895. Johannesburg, South Africa (2013)
10. Zhao, G., di Prisco, M., Vandewalle, L.: Experimental investigation on uniaxial tensile creep behavior of cracked steel fibre reinforced concrete. Mater. Struct. **48**(10), 3173–3185 (2015)
11. Gettu, R.: Creep in the post-cracked region. Final report of sub-task 5.2, Test and design methods of steel fibre reinforced concrete, Brite-Euram project BRPR-CT98-813 (2002)
12. Buratti, N., Mazzotti, C.: Effects of different types and dosages of fibres on the long term behaviour of fibre-reinforced self-compacting concrete. In: Proceedings of 8th RILEM Inertnational Symposium on Fibre Reinforced Concrete, RILEM PRO88, pp. 715–725. Guimarães, Portugal (2012)
13. Mackay, J., Trottier, J. F.: Post-crack behavior of steel and synthetic FRC under flexural creep. In: Shotcrete, Proceedings 2nd International Conference. on Engineering, pp. 183–192. Cairns, Australia (2004)
14. García-Taengua, E., Arango, S., Martí-Vargas, J.R., Serna, P.: Flexural creep of steel fibre reinforced concrete in the cracked state. Constr. Building Mater. **65**, 321–329 (2014)
15. Cochrane, J. T.: Flexural Creep Behaviour of Fibre Reinforced Concrete under High Temperatures, Master of Applied Science Thesis, Dalhousie University, Canada (2003)
16. Buratti, N., Mazzotti, C., Savoia, M., Rossi, P.: Temperature and loading level effect on the long-term behaviour of MSFRC and SFRC. In: Fibre Concrete 2011, Proceedings of 6th International Conference Fibre Concrete, pp. 101-110. Prague, Czech Republic (2011)
17. Kusterle, W., Flexural creep tests on beams-8 years of experience with steel and synthetic fibre. In: FRC-Creep 2016, Proceedings of International RILEM Workshop on Creep Behaviour in Cracked Sections of Fibre Reinforced Concrete, Valencia, Spain (2016)
18. Markovic, I.: High-performance hybrid-fibre concrete: development and utilisation Doctoral dissertation, TU Delft, Delft University of Technology (2006)
19. Bernard, E.S.: Influence of fibre type on creep deformation of cracked fibre-reinforced Shotcrete Panels. ACI Mater. J. **107**(5), 474–480 (2010)
20. Isla Calderón, F. A., Ruano Sandoval, G., Luccioni, B. M.: Analysis of steel fiber pull-out. Experimental study. Const. Building Mater. **100**, 183–193 (2015)
21. Nieuwoudt, P. D., Boshoff, W. P.: Time-dependent behaviour of cracked steel fibre reinforced concrete: on the single fibre and macro level. In: Serna, P. et al. (eds.) FRC-Creep 2016, Proceedings of International RILEM Workshop on Creep Behaviour in Cracked Sections of Fibre Reinforced Concrete, Valencia, Spain (2016)
22. Zhang, X.: Fundamentals of Fibre Science. DEStech Publications, Lancaster, Pennsylvania, USA (2014)
23. Nishiwaki, T., Kwon, S., Otaki, H., Igarashi, G., Shaikh, F.U.A., Fantilli, A.P.: Experimental study on time-dependent behavior of cracked UHP-FRCC under sustained loads. In: Serna, P. et al. (eds.) FRC-Creep 2016, Proceedings International RILEM Workshop on Creep Behaviour in Cracked Sections of Fibre Reinforced Concrete, Valencia, Spain (2016)
24. Daviau-Desnoyers, D., Charron, J.P., Massicotte, B., Rossi, P., Tailhan, J.-L.: Characterization of macrocrack propagation under sustained loading in steel fibre reinforced concrete. Mater. Struct. (2015). doi:10.1617/s11527-015-0552-3. 14 p
25. Garcia-Taengua, E., Llano-Torre, A., Marti-Vargas, J.R., and Serna, P.: Effect of residual strength parameters on FRC flexural creep: multivariate analysis. In: Serna, P. et al. (eds.) FRC-Creep 2016, Proceedings of International Rilem Workshop on Creep Behaviour In Cracked Sections of Fibre Reinforced Concrete, Valencia, Spain (2016)

Creep Testing Methodologies and Results Interpretation

Nicola Buratti and Claudio Mazzotti

Abstract This paper presents a literature review on the main testing methodologies used to investigate the long-term behaviour of cracked FRC elements. Various tests methods such as pull-out tests, uniaxial tension tests, beam bending tests, and plate bending tests are illustrated and discussed. The paper originates from a round table held during the FRC-CREEP 2016 workshop.

Keywords FRC · Creep · Testing · Pull-out · Uniaxial tension · Bending · Plate

1 Introduction

Many studies have contributed to a better characterization of short-term mechanical performances of FRCs and have led to the definition minimum performance requirements and design guidelines. On the other hand, a proper knowledge of the long–term behaviour of SFRCs (Steel Fibre Reinforced Concretes) and MSFRCs (Macro-synthetic Fibre Reinforced Concretes) has not yet been achieved.

Kurtz and Balaguru [1] tested the long-term performance of cracked beams made of concrete reinforced with polypropylene and nylon short-fibres. Creep failure occurred when the stress level was higher than a certain percentage of the failure load under short-term monotonic testing, measured using the Average Residual Strength (ARS) as defined by ASTM C1399 [2]. Bernard [3] investigated the time dependent behaviour of cracked FRC round panels reinforced with either steel or MS fibres. The load applied during long-term tests was defined according to the actual residual tensile-strength measured in the pre-cracking tests. Post-crack creep coefficients were relatively insensitive to load ratio for the SFRC and for one of the two MSFRCs while for the second MSFRC the creep coefficient was sensitive to the load ratio. MacKay and Trottier [4] described the results of experimental tests comparing the behaviour of one SFRC and one MSFRC under long terms loads.

N. Buratti (✉) · C. Mazzotti
DICAM—Structural Engineering, University of Bologna, Bologna, Italy
e-mail: nicola.buratti@unibo.it

P. Serna et al. (eds.), *Creep Behaviour in Cracked Sections of Fibre Reinforced Concrete*, RILEM Bookseries 14,
DOI 10.1007/978-94-024-1001-3_2

They concluded that, at similar loading levels, cracked MSFRCs could experience creep coefficients larger than SFRCs by a factor two. Kusterle [5] tested one SFRC and three different MSFRCs. For each mix six beams were cast and tested in four-point bending. The beams were pre-cracked up to a mid-span deflection of 1.75 mm. A sustained load ranging from 50 to 60 % of the strength at 1.75 mm was applied. Kusterle concluded that MSFRCs had large long-term deformations and that a maximum creep load ratio of 50 % seemed to be the maximum for obtaining good long-term performance. SFRCS were able to sustain larger loads (60 %). Failures were observed when the load level was increased to 75–80 %. Zerbino and Barragán [6] studied the creep behaviour of SFRC cracked beams subjected to long-term loading. The beams were pre-cracked up to crack openings spanning from 0.2 and 3.5 mm. For small crack-openings at the beginning of long-term tests stable responses were obtained during 18 months, even when applying stress levels equal to the final stress level reached at the end of the initial cracking tests. A stable response could be observed for a pre-crack of 0.5 mm. However, for load ratios of 0.96 relatively high crack-opening rates were found, indicating the possibility of the initiation of creep failure. When the loads were further increased, a quick failure was observed in these cases. When creep rupture took place, a three-stage creep response was observed. García-Tengua et al. [7] tested 31 SFRC specimens in four-point bending [8] in order to investigate the effects of various parameters on creep in cracked conditions by means of multiple linear regression. They concluded that the load-ratio had an effect on flexural creep response and that the extent of this effect depends on fibre slenderness and fibre dosage. Zhao et al. [9, 10] carried out an experimental program to investigate the long-term behaviour of SFRCS under uniaxial tensile loads by testing cylindrical specimens, pre-cracked up to either 0.05 or 0.2 mm crack openings. The time-dependent crack opening observed was almost at the same level of instantaneous crack opening after 3 months loading at around 30 % of cracking strength. They also concluded that the damage due to debonding at the fibre/matrix interface was not increasing with creep deformation at the loading level of 30 %, even though the irreversible part almost doubled during the creep loading. Babafemi and Boshoff [11] investigated the time-dependent behaviour of a MSFRC under long-term uniaxial tensile loading. Prismatic specimens were pre-cracked up to 0.5 mm using a displacement control machine. Babafemi and Boshoff observed that the MSFRC showed significant crack widening over time under sustained uniaxial tensile loads. Even at loads as low as 30 % of the post-peak resistance, the time-dependent crack widening did not stabilize after 8 months. Tensile creep failure occurred within 10 days for specimens loaded at 60 % of the post-peak resistance and within less than a day for a 70 % loaded specimen. Average fibre counts on the cracked face of MSFRC were found to influence the time-dependent behaviour. Higher fibre counts resulted in lower time-dependent CMOD and vice versa. Babafemi and Boshoff also performed single fibre long-term pull-out tests observing that specimens loaded at 50 % of the quasi-static capacity pulled out over time. Time-dependent crack widening under sustained loading was identified to be caused by two mechanisms: time-dependent fibre pull-out and time-dependent fibre creep. Buratti and Mazzotti found that temperature influences the creep deformation rates in particular on MSFRCs [12].

This paper presents the most widely experimental techniques used to evaluate creep deformations in cracked FRC sections, as discussed during a round table held during the 1st International RILEM Workshop on creep behaviour in cracked section of Fibre Reinforced Concrete, Valencia, Spain.

2 Pull-Out Tests

Pull-out tests have been used by various researchers in order to understand the bond between fibres and concrete. Most of the studies available in the literature concern the short term behaviour of steel fibres [13–19] but it is possible to find also studies on polymeric fibres [20–24].

To the authors knowledge the only experimental results published based on long-term pull out tests are those by Babafemi and Boshoff [11] and Nieuwoudt et al. [25], who used test setup in which the sustained load was applied to single fibres bonded in concrete using free hanging weights. The pull-out displacement was measured optically by taking digital images with a microscope.

Pull-out tests are attractive because they might allow a better understanding of the bond behaviour, even though using information on the long-term behaviour obtained from these tests in order to predict the behaviour of cross sections might be complicated by random fibre orientation and anchorage lengths.

3 Uniaxial Tension Tests

Van Mier and van Vliet [26] provide an overview on uniaxial tension tests procedures on quasi-brittle materials in general. The heterogeneity of these materials leads to various issues such as the need of large specimens, and the development of secondary flexural moments, causing redistribution of stresses. The experience of uniaxial tension tests on FRCs is in general limited. Plizzari et al. [27, 28] used uniaxial tension tests in order to investigate the behaviour of Steel Fibre Reinforced Concretes (SFRCs) under cyclic loads. Li et al. [29] investigated the uniaxial behaviour of SFRC and Macro Synthetic Fibre Reinforced Concrete (MSFRC) specimens with fibre volume fractions from 2 to 6 %. RILEM TC-162 [30] issued guidelines for uniaxial tension tests based on notched cylindrical specimens and fixed end plates. These guidelines were then used in a round robin test program [31, 32] after which it was concluded that significant differences resulted from different testing labs. The authors suggested that they could be due to different stiffness in the setups. A large variability of the results was observed and was suggested that it was due to material variation, even if it was not possible to analyse whether the small cross section size of the cylinders played an important role. In spite of some difficulties encountered during test executions the authors concluded that the uniaxial test was robust. Using the same experimental setup Barragán et al. [33] performed a parametric study on specimens containing steel fibres and analysed the effects of various parameters such as the depth of the notch and the slenderness of the cylinders. They also investigated the effect of fibre orientation by considering cast and cored specimens, these later being cored from prims in different directions with respect to the prims axis. They concluded that specimens are to be cored in order to obtain representative fibre orientations, especially if they are to be compared with bending tests. This was confirmed also by Zhao et al. [10] and Buratti and Mazzotti [34].

Zerguini and Rossi [35] carried out uniaxial tension tests on SFRCS considering notched cylindrical specimens with different diameters (68, 100, and 150 mm) in order to investigate the sensitivity of test results to this parameter. Fibre dosages from 54 to 100 kg/m^3 where used. The authors concluded that no significant dependence of either the average post-cracking energy in uniaxial tension or the dispersion relative to this characteristic with respect to specimen dimension could be identified, although it should be noticed that high fibre dosages were used. Sorelli et al. [36] carried out tests with freely rotating platens on prismatic specimens.

The literature on long-terms uniaxial tension tests is even more limited. Zhao et al. [10] performed uniaxial creep tests on notched cylinders. The setup used was composed of stiff steel plates epoxied at the ends of the cylinders which were connected by three loading bars used both to apply and to measure (using strain gauges) the tension force. The load was then applied by turning the fastening bolts on the loading bars. During the pre-cracking phase, the three loading bolds were fastened in such a way that the crack opening, which was measured using LVDTs in three positions at 120° around the circumference, was as uniform as possible. After pre-cracking the specimens were then tested under long-term loads using a lever based system. In this phase of the tests the rotation of the specimen ends was not blocked. Babafemi and Boshoff [11] carried out uniaxial tension tests on notched MSFRC prismatic specimens. Pre cracking was carried out in displacement control and then the specimens were loaded using a lever system. Buratti and Mazzotti [34] proposed a test procedure on notched cylinders on which the specimen ends are allowed to rotate, both in the pre-cracking and in the long-term test phase, by means of a spherical joint. During the pre-cracking phase, the authors identified specimens with anomalous rotations (e.g. parts in compression) which were then not used for long-term tests [37]. Long-term tests were then carried out on a chain of three specimens loaded using a lever frame (Fig. 1).

In all the cases cited above the typical testing procedure can be summarized as follows (see Fig. 2). The specimen if first pre-cracked up to a defined crack-opening (O-A) at which the residual strength f_R is measured. The average crack opening is normally considered to quantify the crack opening. Then the specimen it is unloaded (A-B) and transferred to the long-term testing frame. Since pre-cracking is normally carried out at very low crack opening rates, and therefore creep deformations develop during this stage, delayed deformations are in part recovered in the

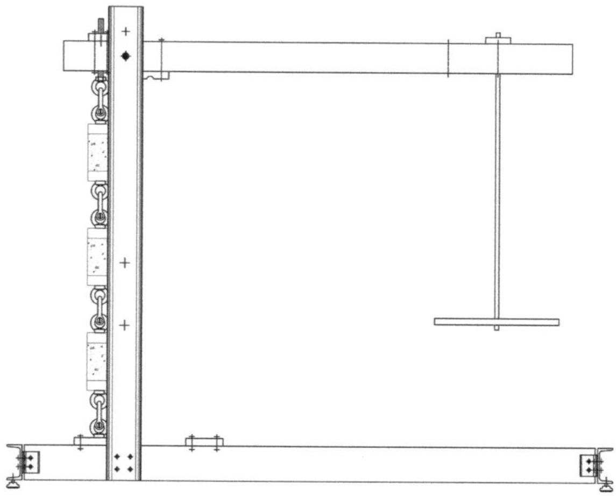

Fig. 1 Experimental setup used by the authors for long-term uniaxial tension tests [37]

unloaded state (B-C). There is yet no agreement in the scientific community on the significance of these deformations and on whether they should be monitored. The specimen is then reloaded up to a fraction of the residual stress measured during pre-cracking (C-D), $\sigma_C = \alpha f_R$, where α indicates the creep load ratio. The time t_L in Fig. 2 indicates the reloading time which should be as short as possible, in order to limit the interference between instantaneous and creep deformations. During the long-term tests deformations will increase at a constant load (D-E). At the end of the long term test the specimen is unloaded (E-F) and part of the long-term deformation is recovered (F-G). Finally, the specimen might be reloaded to failure in a short term test (G-H). It is worth noticing that tertiary creep (leading to failure) might be observed during the long-term tests. These tests are normally carried out in controlled environmental conditions.

In the authors' opinion, even though testing protocols exist for short term tests [38], long term tests present some specific problems. In fact, during long-term tests it is in general not possible to block the rotation of the specimen ends. If specimens are pre-cracked using fixed ends an inconsistent behaviour might be observed during long-terms tests, because of different secondary moments. Furthermore, very large rotations can occur during the long-term tests, leading in some case to compression forces in a portion part of the cracked section. This is more likely to occur when the number of fibres crossing the crack surfaces is limited [34, 37]. Buratti and Mazzotti also observed that uniaxial tension tests, because of the small cross-section size of the specimens, tend to exhibit, on FRCs with low fibre dosages, a scatter of the results which is much higher that bending tests.

Uniaxial tension test, even if more complicated to carry out than bending tests (see Sect. 4), are, to the Authors' opinion an interesting option because, if properly executed, they allow to study the average long term behaviour in tension and therefore their interpretation is more straightforward with respect to bending tests (see Sect. 4).

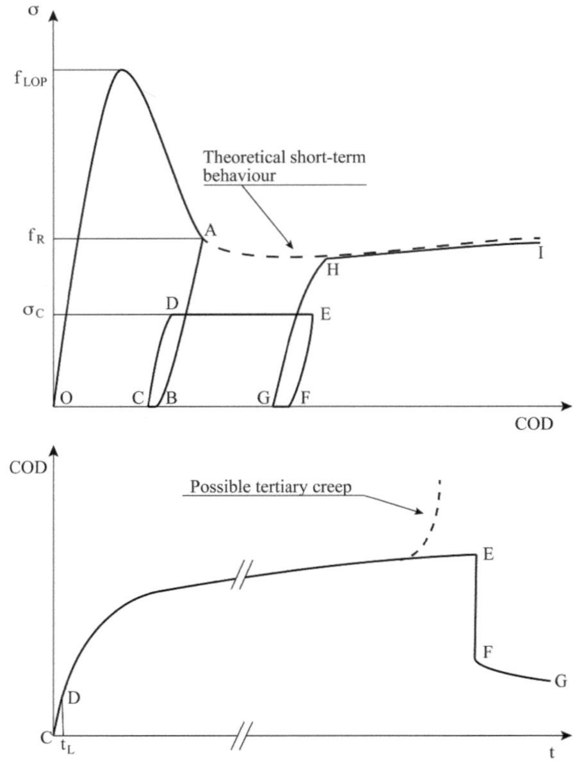

Fig. 2 Typical behaviour observed, in a softening FRC, in long-term tests in cracked conditions, in terms of nominal-stress versus crack opening (*top panel*) and crack opening versus time. COD indicates the general crack opening displacement and might correspond to either CMOD, CTOD or other parameters depending on the test setup

4 Bending Tests

4.1 Test Setup

Bending tests are by far the most widely used in order to evaluate creep deformations in cracked conditions. The testing procedure is normally similar to the one described in Sect. 3 with reference to uniaxial tests (see Fig. 2). The main difference is, of course, the crack opening or deformation measure being used in the pre-cracking and long-terms testing phases (e.g. CMOD, CTOD or mid-span displacement). Typical setups make use of lever systems in order to maintain constant loads for the whole duration of the tests [8]. Figure 3 shows the setup used by the Authors and derived from [8]. Daviau-Desnoyers et al. [39] adapted a hydraulic loading frame for creep tests in order to perform long-term bending tests. Kurtz and Balaguru [1] used a setup in which clamped specimens are loaded by a lever in a cantilever behaviour. Buratti and Mazzotti [12] carried out tests using 200 cm long beams with free hanging weights in a four-point bending setup. In the literature it is possible to find examples of tests on individual specimens and of tests on piles of specimens (normally 3) stacked one on top of each other (e.g. [5, 8, 40]). In this case the load on the specimens is obviously larger at the bottom of the pile, but

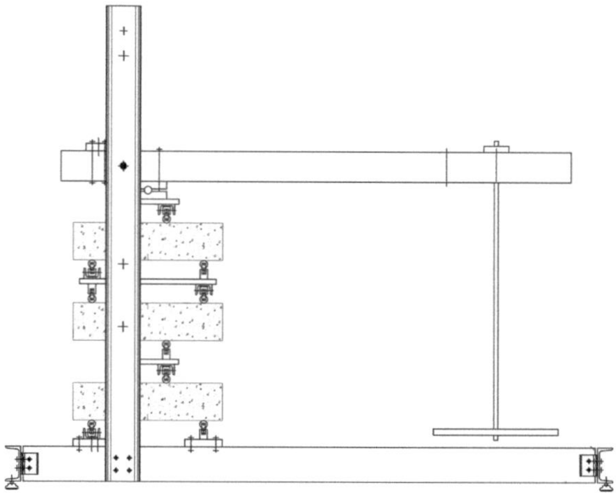

Fig. 3 Experimental setup used by the authors for long-term bending test [37]

often specimens can be ordered in such a way that the ratio of applied force versus residual strength at pre-cracking is constant. This is typically easier on MSFRC specimens because of smaller variability in the post peak behaviour.

In most of the tests published notched specimens were used [4, 6–8, 12, 40, 41] but examples of tests on un-notched specimens can be found in the literature [1, 5]. The most typical setup adopted while testing staked specimens is four-point bending, for obvious stability advantages. One of the shortcoming of this setup is that it is not consistent with the short-term testing procedure proposed by RILEM [42] which is typically used in the pre-cracking stage. However, Zerbino et al. [41] showed that use of three or four-point loading configuration in bending does not affect creep tests results in terms of the COD rate or the stress levels where stable creep behaviour takes place.

In the literature concerning long-term bending test there is not yet agreement on the extent to which some parameters might influence the test results. The Creep load ratio (see Sect. 3) is very often set to 50 % of the residual strength measured at the end of pre-cracking. Many results seem to indicate that creep load ratios between 60 and 70 % might represent an upper limit to the long-term load (see Sect. 1). Defining the load ratio as a percentage of the residual strength at one point might be non-representative for FRCs with strong softening or hardening behaviour. In these cases, in fact, the ratio of creep load versus actual residual strength during the long-term tests might change significantly because of the long-term crack opening. Other criteria were used by some researchers (e.g. [1]) but they are very limited in the literature on the topic. Another important parameter is the crack opening used in the pre-cracking stage, which controls the damage level at the beginning of the long-term tests. According to the previous considerations on the load ratio it might be important to consider, together with the serviceability behaviour, the shape of the stress crack opening curve in the post-peak region in order to define this parameter. Other factors that affect the long term behaviour are environmental conditions, i.e. temperature and relative humidity. Buratti and Mazzotti [12] found that moderate temperature variations could trigger tertiary creep failures in MSFRC specimens. Another factor on which there is no wide agreement in the literature is the minimum test duration. Zerbino et al. [41] suggested that the duration should be around 90 days and proposed to use the 30–90 days crack opening rate, defined as the secant of the COD versus time curve from 30 to 90 days, as synthetic parameter for characterizing the behaviour of different FRCs. Comparing the durations published bending tests of FRCs with typical creep tests in compression these latter are much longer, if fact they have normally a duration of at least 6–12 months. Test duration might be critical because Kusterle observed tertiary creep failures after years of testing [43]. Zerbino et al. [41] found that performing loading-unloading cycles does not contribute to the reduction of testing time. Daviau-Desnoyers et al. [39] performed loading-unloading cycles during long-term tests in order to evaluate the progression of damage based on the observed compliance of the specimens.

Finally, another point on which different authors adopted different approaches is related to the deformation parameters to be measured during the tests. Typical examples are CMOD, CTOD and mid-span deflection. Relationships among those parameters (in particular crack opening and deflection) for long-term loads are not yet fully defined and therefore it might be complicated to compare tests in which different parameters were used. As an example, the creep coefficient in terms of mid-span deflection is normally larger than the one in terms of CMOD [12] because the former deformation parameter is affected not only by the delayed behaviour of the cracked cross-section but also by the creep of the un-cracked parts of the specimen.

4.2 Interpretation of Results

Most of the research results published until now provided a phenomenological description of the long-term behaviour of FRC elements in cracked conditions and identified some significant parameters. Analytical models were used to provide a description of the long-term behaviour ad it was shown that functions that are normally used to describe creep in compression can properly represent creep in bending [3, 44].

Detailed interpretation of results from long-term bending tests is still an open problem. In fact, the behaviour of the specimen, either in terms of CMOD or in terms of mid-span deflection, is affected by both the creep concrete matrix and the creep of the cracked part. This latter is the result on different phenomena that interact and whose individual contributions have not been yet clearly defined: creep of the concrete matrix in tension, creep of the fibre-to-concrete bond, micro-cracking [45], and, as far as MS fibres are concerned, creep of the fibres [46]. Concerning this latter contribution, it is worth noticing that currently there are no minimum requirements for MS fibres in terms of creep deformation.

In order to introduce creep in design it is mandatory to separate the tension behaviour from the one in compression. A viable approach seems to be constituted by inverse analysis. Buratti and Mazzotti [47], for example, proposed to derive compression creep form specific tests and to use a fibre based-model to provide a phenomenological description of the creep in tensions (using an equivalent continuum model). This model was then used to estimate creep in tension using an inverse analysis procedure on long-term flexural data. The model is being further extended using data from uniaxial tension tests for defining the creep model for the portion of the cross section in tension.

5 Plate Tests

Long-terms tests on circular plates were proposed by Bernard [3] with the aim of characterizing creep deformations in cracked conditions for fibre reinforced shotcrete elements. Recently long term tests on square panes have been proposed by Larive et al. [48]. The interpretation of the results of these tests presents complications similar to those discussed in Sect. 4.2, which in this case are increased by the static redundancy of the specimens and by the presence of multiple cracks. Furthermore, more complex experimental setups are typically required. Alternative testing procedures proposed by Ciancio et al. are being investigated by the Authors to this aim [49, 50].

6 Conclusions

A significant deal of research has been focused, in recent years, on the characterization the long-term behaviour of FRC elements in cracked conditions. Even if different researchers are converging on similar testing protocols the following main conclusions can be drawn from the literature review presented by the present paper:

- standard testing protocols, and relationship among main deformation parameters, need to be developed in order to allow and easier comparison of results and to facilitate discussion among researchers.
- The extent to which parameters such as creep load ratio, crack opening at the beginning of the long-term test, temperature and relative humidity influence the behaviour of FRCs has not yet been fully defined.
- Chemical, physical and mechanical phenomena governing creep of FRCs in cracked conditions have no yet been fully understood.
- The conditions for the onset of tertiary creep have not been yet fully understood.
- Inverse analysis procedures are most probably required for the interpretation of long-term bending tests.

– Prediction models for FRC creep in cracked conditions and in particular for tensile creep are not yet available but are needed if these phenomena are to be included in design.

Acknowledgments The Authors would like to acknowledge the contribution of all the members of the RILEM TC 261-CCF for their important contributions to the round table from which this paper was derived.

References

1. Kurtz, S., Balaguru, P.: Postcrack creep of polymeric fiber-reinforced concrete in flexure. Cem. Concr. Res. **30**(2), 183–190 (2000)
2. American Society for Testing and Materials (ASTM) International, 'C1399 Standard Test Method for Obtaining Average Residual-Strength of Fiber-Reinforced Concrete', 2010
3. Bernard, E.S.: Creep of cracked fibre reinforced shotcrete panels. In: Shotcrete: More Engineering Developments, pp. 47–57. Taylor & Francis, London (2004)
4. MacKay, J., Trottier, J.F.: Post-crack creep behavior of steel and synthetic FRC under flexural loading. In: Shotcrete: More Engineering Developments, pp. 183–192. Taylor & Francis, London (2004)
5. Kusterle, W.: Viscous material behavior of solids-creep of polymer fiber reinforced concrete. 5th Central European Conference on Concrete Engineering, pp. 95–99 (2009)
6. Zerbino, R.L., Barragán, B.E.: Long-term behavior of cracked steel fiber-reinforced concrete beams under sustained loading. ACI Mater. J. **109**(2), 215–224 (2012)
7. García-Taengua, E., Arango, S., Martí-Vargas, J.R., Serna, P.: Flexural creep of steel fiber reinforced concrete in the cracked state. Constr. Build. Mater. **65**, 321–329 (2014)
8. Arango, S.E., Serna, P., Martí-Vargas, J.R., García-Taengua, E.: A test method to characterize flexural creep behaviour of pre-cracked FRC specimens. Exp. Mech. **52**(8), 1067–1078 (2012)
9. Zhao, G., Di Prisco, M., Vandewalle, L.: Experimental research and numerical simulation of post-crack creep behavior of SFRC loaded in tension. Mechanics and Physics of Creep, Shrinkage, and Durability of Concrete, Proceedings of the Ninth International Conference on Creep, Shrinkage, and Durability Mechanics (CONCREEP-9), pp. 340–347, Cambridge, Massachusetts (2013)
10. Zhao, G., di Prisco, M., Vandewalle, L.: Experimental investigation on uniaxial tensile creep behavior of cracked steel fiber reinforced concrete. Mater. Struct. **48**(10), 3173–3185 (2015)
11. Babafemi, A.J., Boshoff, W.P.: Tensile creep of macro-synthetic fibre reinforced concrete (MSFRC) under uni-axial tensile loading. Cement Concr. Compos. **55**, 62–69 (2015)
12. Buratti, N., Mazzotti, C.: Experimental tests on the effect of temperature on the long-term behaviour of macrosynthetic Fibre Reinforced Concretes. Constr. Build. Mater. **95**, 133–142 (2015)
13. Naaman, A.E., Namur, G.G., Alwan, J.M., Najm, H.S.: Fiber pullout and bond slip. II: experimental validation. J. Struct. Eng. **117**(9), 2791–2800 (1991)
14. Robins, P.J., Austin, S.A., Jones, P.A.: Pull-out behaviour of hooked steel fibres. Mater. Struct. **35**(7), 434–442 (2002)
15. Cunha, V.M.C.F., Barros, J.A.O., Sena-Cruz, J.M.: Pullout behavior of steel fibers in self-compacting concrete. J. Mater. Civ. Eng. **22**(1), 1–9 (2009)
16. Breitenbücher, R., Meschke, G., Song, F., Zhan, Y.: Experimental, analytical and numerical analysis of the pullout behaviour of steel fibres considering different fibre types, inclinations and concrete strengths. Struct. Concr. **15**(2), 126–135 (2014)

17. Zīle, E., Zīle, O.: Effect of the fiber geometry on the pullout response of mechanically deformed steel fibers. Cem. Concr. Res. **44**, 18–24 (2013)
18. Choi, W.C., Jang, S.J., Do Yun, H.: Interface bond characterization between fiber and cementitious matrix. Int. J. Polym. Sci. **2015**, 1–11 (2015)
19. Isla, F., Ruano, G., Luccioni, B.: Analysis of steel fibers pull-out. Experimental study. Constr. Build. Mater. **100**, 183–193 (2015)
20. Li, V.C., Wang, Y., Backer, S.: Effect of inclining angle, bundling and surface treatment on synthetic fibre pull-out from a cement matrix. Composites **21**(2), 132–140 (1990)
21. Singh, S., Shukla, A., Brown, R.: Pullout behavior of polypropylene fibers from cementitious matrix. Cem. Concr. Res. **34**(10), 1919–1925 (2004)
22. Lee, S.C., Shin, K.J., Oh, B.H.: Cyclic pull-out test of single PVA fibers in cementitious matrix. J. Compos. Mater. **45**(26), 2765–2772 (2011)
23. Di Maida, P., Radi, E., Sciancalepore, C., Bondioli, F.: Pullout behavior of polypropylene macro-synthetic fibers treated with nano-silica. Constr. Build. Mater. **82**, 39–44 (2015)
24. Alberti, M.G., Enfedaque, A., Gálvez, J.C., Ferreras, A.: Pull-out behaviour and interface critical parameters of polyolefin fibres embedded in mortar and self-compacting concrete matrixes. Constr. Build. Mater. **112**, 607–622 (2016)
25. Nieuwoudt, P.D., Boshoff, W.P.: Modelling the time-dependent pull-out behaviour of hooked steel fibres. In: Proceedings of CONCREEP 10—Mechanics and Physics of Creep, Shrinkage and Durability of Concrete and Concrete Structures, pp. 1333–1339, Vienna, Austria, 21–23 Sept 2014
26. van Mier, J.G.M., van Vliet, M.R.A.: Uniaxial tension test for the determination of fracture parameters of concrete: state of the art. Eng. Fract. Mech. **69**(2), 235–247 (2002)
27. Plizzari, G.A., Cangiano, S., Alleruzzo, S.: The fatigue behaviour of cracked concrete. Fatigue Fract. Eng. Mater. Struct. **20**(8), 1195–1206 (1997)
28. Plizzari, G.A., Cangiano, S., Cere, N.: Postpeak behavior of fiber-reinforced concrete under cyclic tensile loads. ACI Mater. J. **97**(2), 182–192 (2000)
29. Li, Z., Li, F., Chang, T.-Y.P., Mai, Y.-W.: Uniaxial tensile behavior of concrete reinforced with randomly distributed short fibers. ACI Mater. J. **95**(5), 564–574 (1998)
30. Vandewalle, L., Nemegeer, D., Balazs, L., Barr, B., Bartos, P., Banthia, N., Brandt, A., Criswell, M., Denarie, F., Di Prisco, M., Falkner, H., Gettu, R., Gopalaratnam, V., Groth, P., et al.: RILEM TC 162-TDF: test and design methods for steel fibre reinforced concrete—Uni-axial tension test for steel fibre reinforced concrete. Mater. Struct. **34**(235), 3–6 (2001)
31. Barr, B., Lee, M., Barragán, B., Dupont, D., Gettu, R., Olesen, J.F., Stang, H., Vandewalle, L.: Round-robin analysis of the RILEM TC 162-TDF uni-axial tensile test: part 1. Mater. Struct. **36**(4), 265–274 (2003)
32. Barr, B., Lee, M., Barragán, B., Dupont, D., Gettu, R., Olesen, J.F., Stang, H., Vandewalle, L.: Round-robin analysis of the RILEM TC 162-TDF uni-axial tensile test: part 2. Mater. Struct. **36**(4), 275–280 (2003)
33. Barragán, B.E., Gettu, R., Martín, M.A., Zerbino, R.L.: Uniaxial tension test for steel fibre reinforced concrete—a parametric study. Cement Concr. Compos. **25**(7), 767–777 (2003)
34. Buratti, N., Mazzotti, C.: Uniaxial tension tests on macrosynthetic fibre reinforced concretes. In: Proceedings of the 9th Rilem International Symposium on Fiber Reinforced Concrete (BEFIB), Vancouver, Canada, 19–21 Sept 2016
35. Zerguini, A., Rossi, P.: Post-cracking behaviour in uniaxial tension of metallic fiber-reinforced concrete (MFRC): experimental study of scale effects. Bulletin des Laboratoires des Ponts et Chaussées **242**, 67–75 (2003)
36. Sorelli, L.G., Meda, A., Plizzari, G.A.: Bending and uniaxial tensile tests on concrete reinforced with hybrid steel fibers. J. Mater. Civ. Eng. **17**(5), 519–527 (2005)
37. Buratti, N., Mazzotti, C.: Experimental tests on the long-term behaviour of SFRC and MSFRC in bending and direct tension. In: Proceedings of the 9th Rilem International Symposium on Fiber Reinforced Concrete (BEFIB), Vancouver, Canada, 19–21 Sept 2016

38. RILEM TC 162-TDF: Test and design methods for steel fibre reinforced concrete. Uni-axial tension test for steel fibre reinforced concrete. Recommendations. Mater. Struct. **34**(1), 3–6 (2001)
39. Daviau-Desnoyers, D.L., Charron, J.-P., Massicotte, B., Rossi, P., Tailhan, J.-L.: Creep behaviour of a SFRC under service and ultimate bending loads. In: Proceedings of the International RILEM Workshop on creep behaviour in cracked section of Fibre Reinforced Concrete, Valencia Spain, 9–10 March 2016
40. Buratti, N., Mazzotti, C., Savoia, M.: Long-term behavior of cracked SFRC elements exposed to chloride solutions. Advances in FRC Durability and Field Applications (ACI SP-280) (2011)
41. Zerbino, R., Monetti, D.H., Giaccio, G.: Creep behaviour of cracked steel and macro-synthetic fibre reinforced concrete. Mater. Struct. **49**(8), 3397–3410 (2016)
42. Barr, B., Lee, M., de Place Hansen, E., Dupont, D., Erdem, E., Schaerlaekens, S., Schnütgen, B., Stang, H., Vandewalle, L.: Round-robin analysis of the RILEM TC 162-TDF beam-bending test: part 1—test method evaluation. Mater. Struct. **36**(9), 609–620 (2003)
43. Kusterle, W.: Flexural creep tests on beams: 8 years of experience with steel and synthetic fibres. In: Proceedings of the International RILEM Workshop on creep behaviour in cracked section of Fibre Reinforced Concrete, Valencia Spain, 9–10 March 2016
44. Babafemi, A.J., Boshoff, W.P.: Testing and modelling the creep of cracked macro-synthetic fibre reinforced concrete (MSFRC) under flexural loading. Mater. Struct. **49**(10), 4389–4400 (2016)
45. Nieuwoudt, P.D., Boshoff, W.P.: The time-dependant pull-out behaviour of hooked steel fibres. In: Proceedings of the International RILEM Workshop on creep behaviour in cracked section of Fibre Reinforced Concrete, Valencia Spain, 9–10 March 2016
46. Vrijdaghs, R., di Prisco, M., Vandewalle, L.: Creep deformations of structural polymeric macrofibers. In: Proceedings of the International RILEM Workshop on creep behaviour in cracked section of Fibre Reinforced Concrete, Valencia Spain, 9–10 March 2016
47. Buratti, N., Mazzotti, C.: An experimental and numerical study on the long-term behaviour of cracked fibre-reinforced self-compacting concrete beams. In: Proceedings of the 7th RILEM Symposium on Self-Compacting Concrete, Paris, France, 2–4 Sept 2013
48. Larive, C., Chamoley, D.R., Regnard, A., Pannetier, T., Thuaud, C.: A new testing method dedicated to creep behaviour of fibre reinforced sprayed concrete. In: Proceedings of the International RILEM Workshop on Creep Behaviour in Cracked Section of Fibre Reinforced Concrete, Valencia Spain, 9–10 March 2016
49. Ciancio, D., Mazzotti, C., Buratti, N.: Evaluation of fibre-reinforced concrete fracture energy through tests on notched round determinate panels with different diameters. Constr. Build. Mater. **52**, 86–95 (2014)
50. Ciancio, D., Manca, M., Buratti, N., Mazzotti, C.: Structural and material properties of mini notched round determinate panels. Constr. Build. Mater. **113**, 395–403 (2016)

Part II
Influence of Fibre Type on Creep

Flexural Creep Tests on Beams—8 Years of Experience with Steel and Synthetic Fibres

Wolfgang Kusterle

Abstract The creep of cracked fibre reinforced concrete beams has been investigated by many laboratories. Different testing procedures have been used. Creep-test results prove that cracked beams reinforced by fibres show time dependent deformations and therefore increasing crack width with time. Some cannot sustain a high percentage of post-crack load for an extended period and fail due to creep rupture. Determination of the residual strength of FRC according to the Austrian Guideline Fibre Reinforced Concrete was the starting point for the tests. The beams were first tested under four-point flexural loading up to a deflection of 1.75 mm. The beams were then unloaded and three pre-cracked beams were loaded with a sustained load at various percentages of the load reported at 1.75 mm deflection. The used testing procedure, realized and planned improvements are discussed. Typical creep-time curves of FRC are presented. The load level was increased stepwise. Starting point was 50 % of the residual load at 1.75 mm deflection. At the load level of 60 %, beams with several fibre types started deforming faster and the test ended up in a creep failure. An alarming fact is that beams with some fibre-types may fail after 7 years of loading.

Keywords Structural fibres · Flexural test · Unnotched beams · Creep · Creep rupture

1 Introduction

Creep is defined as a phenomenon in which strain in a solid slowly increases under constant stress. It is a viscous deformation with some delayed elastic deformation included. Creep in post crack stage is of interest for fibre reinforced concrete at

W. Kusterle (✉)
OTH Regensburg, Regensburg University of Applied Sciences, Regensburg, Germany
e-mail: wolfgang@kusterle.net

P. Serna et al. (eds.), *Creep Behaviour in Cracked Sections of Fibre Reinforced Concrete*, RILEM Bookseries 14,
DOI 10.1007/978-94-024-1001-3_3

different load levels and pre-deformation. Data is missing for discrete fibres produced of different materials and with different geometry at low dosage in contact with a cement matrix at different humidity and temperature. Creep in tension of fibre reinforced concrete may be investigated in direct tension tests, in tests on beams and in test on panels. In this paper flexural creep tests on beams are discussed. Tests on beams are usually used for determining the values for design. Therefore it makes sense to use the same or similar testing procedures for creep investigations, too. Direct tensile tests would be more precise, but are much more complicated to perform, there use is limited to research purposes. Tests on panels are more favourable to show the positive effect of fibres, but are one more step from the needed design values. Panels with continuous support along all four sides will have problems with external and internal friction, especially in the case of long term tests at low load level. However, it must be kept in mind, that flexural tests are sensitive to specimen preparation, curing and humidity during testing, resulting in high scatter of the results. The load-deflection graph is instable after the first crack and often influenced by the loading speed and the used testing machine.

2 Literature Review on Flexural Creep Testing on Beams

Proper creep tests are time consuming and expensive, as good equipment is needed for a long time in a controlled climate. There are many influencing factors, which have to be investigated. Several investigations have dealt with the monitoring of creep deformation on beams reinforced with steel and polymer fibres in flexural tensile tests. A literature review is given in Table 1.

Starting point for most tests is the determination of the residual strength of FRC, according to well-known standards. Usually the creep test starts immediately after the residual strength test and in some cases the residual strength test is continued after creep testing (Fig. 1). In many cases the creep test results in creep rupture and no further testing is possible.

While some tests use uncracked beams, most use precracked beams. The crack width varies between 0.2 and 5 mm.

The test rigs used for creep testing often use lever arm loading, sometimes hydraulic or spring type loading. The loading levels differ substantially and refer sometimes to the ultimate flexural strength, mostly to the load at the maximum deflection in residual strength testing.

Table 1 Creep tests on beams reported in literature [1]

Literature	Dimensions (mm) n: notched	Span (mm)	Specification	Deflection	Load level	Single/Multi set up	Load-transfer	Climate
Arango	150 × 150 × 600 n	500 450	EN 14651	0.5 mm CMOD	60 % 80 %	3	Lever arm	20 °C/50 % rH
García-Taengua	150 × 150 × 600 n	450		0.5 mm CMOD	60–80 %	3	Lever arm	20 °C/50 % rH
Barragán Zerbino	150 × 150 × 600 n	500 450	EN 14651	0.5–3.5 mm CMOD	59 % 60 %	3	Lever arm	16–22 °C 22–64 % rH
Bast, Eder Kusterle	150 × 150 × 500	450	obv Guideline	1.75 mm	50 % 60 %	1	Lever arm	Sealed by Aluminium, 20 °C
BEKAERT Lambrechts	150 × 150 × 500	450	4 point	5 mm	50 %	1	Dead load	Not reported
Ratcliffe	Same as Lambrechts		4 point	5 mm	50 %	1	Not reported	room
Buratti	300 × 120 × 2000 n	800	EN 14651	0.2 mm CMOD	60–70 %	1	Dead load	20/60
Savoia	300 × 120 × 2000 n	750	4 point	0.2 mm CMOD	50 %	Same as Buratti	Dead load	20–40 °C/rH not rep.
Chanvillard	150 × 200 × 700 n	600	–	0.3 mm 0.6 mm	72–80 %	3	Jack	20/50
Cochrane MacKay	100 × 100 × 350 n		ASTM C1399	<0.2 mm	20, 40, 60 % of ARS	3	Lever arm	30–60 °C
DIBt	150 × 150 × 700 100 × 100 × 500	600 450	DBV EN 14651	0.3 mm	50 %	Not spec.	Not spec.	20 °C, rH not spec.
Gossla	150 × 150 × 700 150 × 300 × 700	600 600	DAfStB	0.5 mm	40–50 %	3	Weight + cantilever	20/60
Granju	150 × 200 × 700 n	600	–	0.3 mm CMOD	100 cycl. loaded 50–80 %	1	Not reported	Not reported

(continued)

Table 1 (continued)

Literature	Dimensions (mm) n: notched	Span (mm)	Specification	Deflection	Load level	Single/Multi set up	Load-transfer	Climate
Kanstad	120 × 150 × 600 n	450	–	0.2 mm CMOD	50 %	1	Hydr. cylinders	20/50
Kaufmann	150 × 150 × 550 n 150 × 150 × 700	500 600	EN 14651	0.5 mm	25–91 % of first crack	3	Gas cylinders	20/70
Kurtz	100 × 100 × 350	300	ASTM C1399 + modific.	0.75 mm	25–40 %	1	Cantilever	20/50
Larsen	40 × 60 × 600	550	4 point	Not reported Uncracked	35, 50, 60 % of 28 days	Not reported	Not reported	23/65 23/65–95 outdoor
Tan	100 × 125 × 2000 reinforced	1800	6 point	Uncracked	35–80 % of P_u	1	Dead load	Not reported
Theodorakopoulos	t × 100 × 500	450	4 point	Uncracked	18–75 % of P_u	2	Spring	Indoor

Remark Large beam tests are not reported here. See Ref. [1]

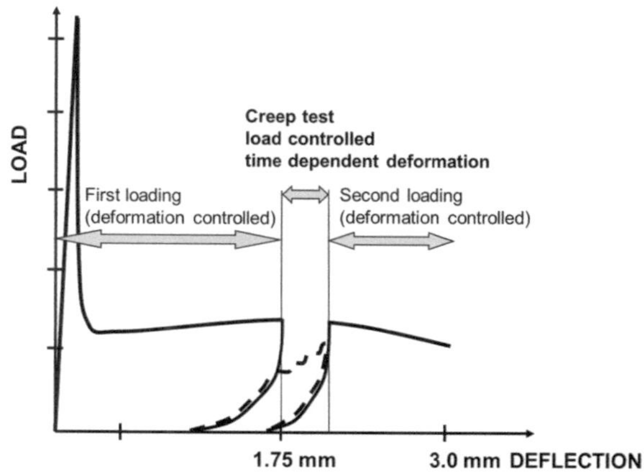

Fig. 1 Idealized plot obtained after complete testing of a beam specimen. Bending test after creep test is seldom performed

2.1 Discussion of Test Procedure

In comparison to panel testing, only a small amount of fibres is oriented in the main stress direction, when using beams. It is therefore necessary to use beams with a large cross section to reduce scatter in results. The span of the beams should not be too small, otherwise you test vault effects instead of real bending.

Notched beams, using centre point loading, have the advantage of direct correlation between load and crack mouth opening displacement, but give higher results (up to 15 %) than tests on unnotched beams in four-point loading. Furthermore the cross section is reduced compared to unnotched beams. As the creep is assessed as a percentage of load at a given deformation, the creep load is effected, too. For practical reasons (stability) many creep tests on notched beams have been performed in four-point loading.

Bending tests on steel fibre reinforced beams often result in different shape of the load-deformation graphs in comparison to beams with synthetic fibres. Both rise and reach a peak when the “first” crack occurs. The graphs of the samples with steel fibres then fall steadily, when the fibres are pulled out. Opposite to this behaviour, samples with synthetic fibres fall sharp before they recover. They reach a second peak before they start falling again (Fig. 2). The deflection, at which they hit a low, often can be 0.1–0.6 mm (tests with 450 mm span). This is in the area of deflection, many creep tests use as starting point. The post crack parameters used for design may be nearly the same for both materials. But, when one compares the behaviour of beams with different fibres, a creep load of the percentage of the residual load at defined deflection is used. The load of beams with steel fibres is much higher in this area of deflection than that of beams with synthetic fibres. The relative elongation of the synthetic fibres is quite low, they are at this period of the test not loaded to the same extend as the much stiffer steel fibres. This may result in misleading conclusions.

In flexural creep test a mixed deformation is monitored resulting from deformations in tensile and compression zones of the cross section. As the creep deformation

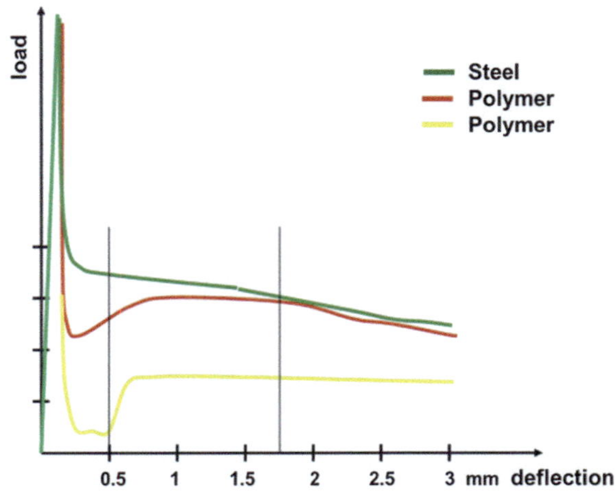

Fig. 2 Typical load-deflection graphs of FRC with steel fibres and synthetic fibres

measured by crack mouth opening or net deflection is a mixture of creep and shrinkage of the concrete and the creep and slip out of the fibres, additional testing or calculations for these properties would be necessary, but are often not available. The environment during testing is mostly kept at ordinary lab conditions, some tests have been done at elevated temperature; other tests cover the beams by aluminium sheets to prevent evaporation. Temperatures higher than 40 °C will clearly increase the deformations, when synthetic fibres are used. The slip of all fibres will be influenced, when the humidity of the concrete matrix changes between water saturated and completely dry conditions.

As quite small deformations are considered, it is of importance that the supports are allowing these deformations with as little friction as possible. As many test rigs are produced as low cost prototypes, these requirements are often not fulfilled completely.

Due to limited access to test rigs and limited space in climate chambers, often several beams are loaded together and the load level is increased stepwise during time. This results in imprecise load levels and lack of data for creep models, which need data from uniform loading at different load levels. Due to lack of time, many tests end too early.

3 Investigations at the OTH

When designing fibre reinforced concrete, in Germany the DAfStb-Guideline “Steel fibre reinforced concrete” [2] is used, whereas in Austria the obv guideline “Fibre reinforced concrete” [3] is used. Both guidelines use unnotched beam testing for the determination of flexural strength and post crack strength. Therefore it makes sense to use the same test procedure for flexural creep testing.

In [3] a test procedure for creep tests is included in the informal part of this guideline. Tests done by the author of this paper have been the basis for these recommendations [4, 5]. This test procedure was followed, when performing the tests described in this paper. First tests started in autumn 2006, and since then the 12 test rigs have been used continuously.

For each type of fibre six beams 150 × 150 × 500 mm were cast and cured under water up to 28 days. Shortly before testing the beams were prepared for testing.

3.1 Testing Post Crack Behaviour

Testing was performed on beams under four-point flexural loading with a span of 450 mm. For the purpose of this investigation the beams were first tested under four-point flexural loading up to a deflection of 1.75 mm. The beams were then unloaded and prepared for creep test.

3.2 Flexural Creep Test

Three pre-cracked beams were loaded with a sustained load at various percentages starting with 50 % of the load reported at 1.75 mm bending deflection. The chronological order of testing is shown in Fig. 1. The limited number of rigs was the reason why different load levels were performed using one specimen. The creep rig is designed similarly to the testing machine used for the flexural test. However, the load is kept at a constant level by simple leverage (Fig. 3). Lubricants were used to minimise friction at the supports. Later on roller bearings were installed at one support.

The test results combine both the creep of concrete in compression and the creep of the fibres in tension. There is also creep deflection due to fibre pull-out over time. The beams have to be kept at constant humidity. Therefore, they were protected from drying out by aluminium sheets. The creep rigs are situated in the basement of the OTH lab with quite uniform climate conditions, but unfortunately not that uniform as in a climate chamber.

Creep load is applied using a balance (Fig. 4). When increasing the load, the necessary dead weight and the leverage arm length has to be calculated. The creep deformation versus time is registered for all specimens in given time intervals. For most samples, the test started with a load level of 50 % of the load at 1.75 mm deflection and was increased after about 90 days to a load level of 60 %, some at later age even to 70 and 80 %.

Mass lost during testing and fibre counting in cross section after the tests were done in a last step for proper interpretation of results.

Fig. 3 The simple, but sufficient stiff test-rig used for these tests

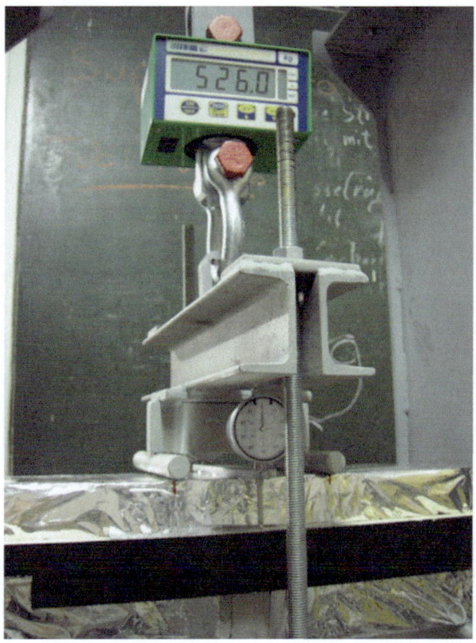

Fig. 4 Applying the exact creep load. *Note* Supporting rollers in this equipment not sufficiently moveable, instead lubricants are used

3.3 Discussion and Possible Improvements

The obv guideline uses a deflection of 1.75 mm for precracking, which is the mean deflection for the determination of ultimate limit state calculation. Most other authors use the deflection for service state calculations (e.g. 0.5 or 1.0 mm). This approach might be more realistic. But the discussion under 2.1 should be kept in mind.

Beams with cross section of 150 × 150 mm and a span of 450 mm will not give such a realistic bending situation as beams with 600 mm span, which are recommended by DAfStb [2].

The influence of humidity, shrinkage and creep of concrete in the compression zone has not been investigated in detail yet. Small temperature changes in our test environment also result in small changes of deformations. Unfortunately it was not possible to test 50 and 60 % loaded specimens in parallel due to limited test facilities. Therefore long-time experience with 50 % is lacking.

The supports were improved by installation of supporting rollers.

Up to now for most tests one mix design has been used. 2015 a RRT started with other mix designs and fibre dosages.

3.4 Concrete Mix Design

A common used mix-design for FRC was used for all tests: 370 kg/m^3 CEM II A-S/42.5 R, w/c = 0.5, 1747 kg/m^3 Danube sand and gravel 4/16.

Consistency was controlled by different dosages of Sika Viscocrete 1020X at 450 mm on the spread table, 28 days compressive strength reached 60 MPa, first peak strength about 6 MPa. The fibre dosage was quite low. About the same amount of fibres, calculated by volume, was added for steel (30.0 kg/m^3) and polymeric fibres (4.5 kg/m^3) in all tests reported here.

For these tests 9 different fibres, representing commonly used types, were used, 8 synthetic macro fibres and one type of a typical steel fibre as Ref. [5]. As structural synthetic macro fibres did improve since 2006, when the tests started, it makes no sense to identify specific fibres. The fibres vary in their material composition and dimensions. The results should mainly give an overview on possible creep deformations with different fibres.

4 Results of Creep Tests

Test results are available now up to more than 3200 days or 8 years. The results show higher deformation for cracked beams reinforced by synthetic fibres at the same load level and therefore wider cracks than for SFRC. Some cannot sustain a

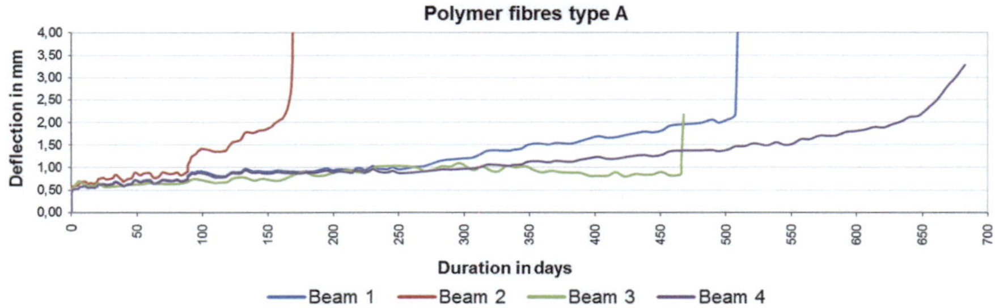

Fig. 5 Creep deformation of the 4 beams with type A polymer fibres, leading to creep rupture after different period of time. *Note* The creep load is 50 % of the load at 1.75 mm up to 90 days, and then increased to 60 %. The absolute values of creep load are not the same for all 4 beams

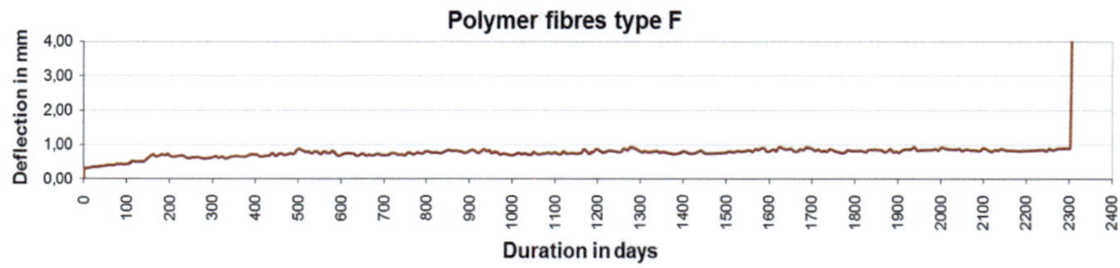

Fig. 6 Creep deformation of one beam with type F polymer fibres, leading to creep rupture after 2300 days. Load level 60 %. Total deflection before tertiary creep very low. The other two beams are still carrying the sustained load after 2694 days

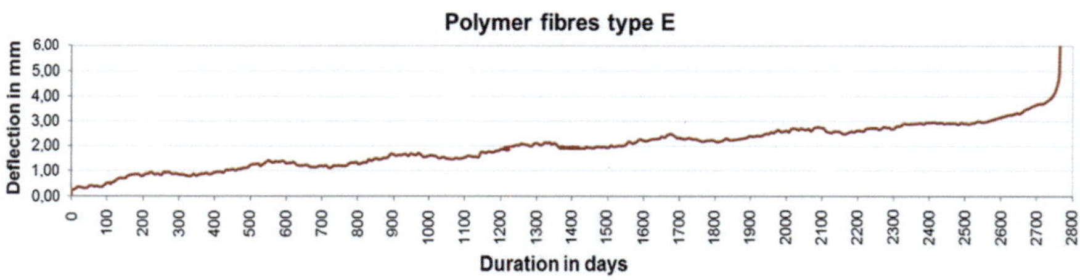

Fig. 7 Creep deformation of one beam with type E polymer fibres, leading to creep rupture after 2770 days. Load level 60 %. Total deflection before tertiary creep is moderate. The other two beams collapsed after 190 and 260 days

high percentage of post-crack load for extended periods and fail due to creep failure.

Some typical results of individual specimens are shown in Figs. 5, 6, 7, 8, 9 and 10.

Figure 5 gives an overview of 4 test results for a typical synthetic fibre in 2006. All 4 samples could not withstand the 60 % of short time load.

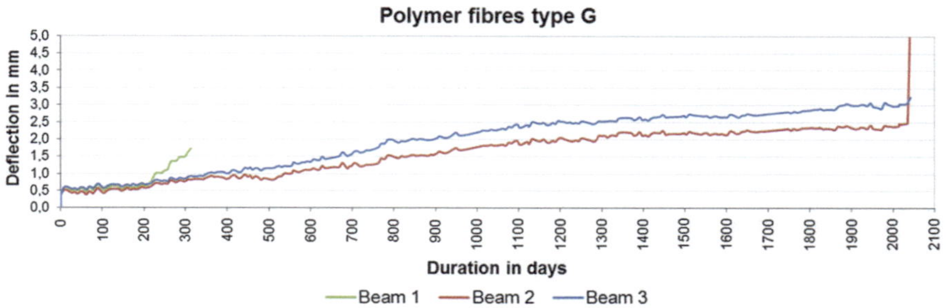

Fig. 8 Creep deformation of three beams with type G polymer fibres: *Beam 1* experienced creep rupture after 313 days. Load level 60 %. *Notice* Absolute sustained load applied was 60 % higher, than for beam 2 and 3. *Beam 2* collapsed after 2041 days. *Beam 3* was still carrying the sustained load. The test was finished at the collapse of beam 2. For both load level 60 % had been changed to 65 % after 2013 days

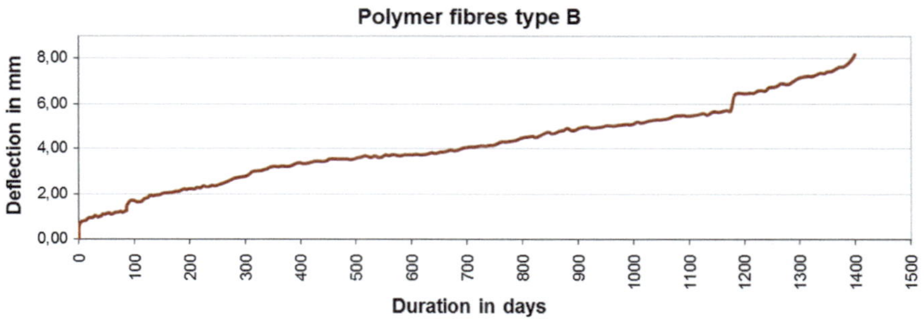

Fig. 9 Creep deformation of one beam with type B polymer fibres, leading to creep rupture after 1400 days. Load level 60 %. Very high total deflection. The other two beams collapsed after 1062 days at 11 mm deflection and 3000 days at 5.5 mm deflection

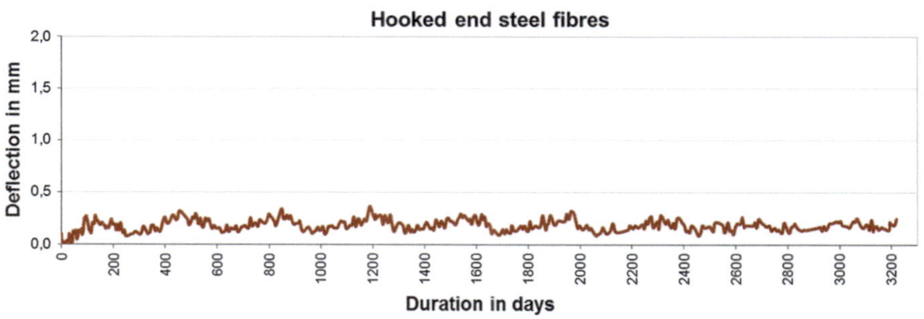

Fig. 10 Creep deformation of one beam with a typical steel fibre, at load level 4 (50 % up to 85 days, 60 % to 3180 days, 70 % to 3190 days, after that the load was increased to 80 %)

While beams with type F fibres (Fig. 6) show low deformation and only one collapsed up to now, the beams with type E fibres (Fig. 7) showed slight, but ever increasing secondary creep, which ended up in creep rupture after 190–2770 days.

Figure 8 shows the influence of the load applied, relatively always 60 %, but absolutely for beam 1 with 60 % difference (10.44 kN instead of 6.42 kN).

The crack width for one specimen reached nearly 10 mm (Fig. 9). But the most alarming fact is that beams with some types of fibres can fail after a long period of testing, e.g. after 6 or 7 years (Figs. 6 and 7). Up to 1.5 years or even longer the graphs of deformation-time still show low rates of creep deformation.

Beams with common steel fibres (low carbon cold drawn wire fibre with hooked ends) are still performing excellently (Fig. 10) at load levels of 60 and 70 %. Note, that the concrete properties will have changed after 3200 days. Beams at 28 days with the same steel fibres under 75 % and up to 85 % of the residual load were tested as well. They failed within minutes due to fibre pull out at this load level.

4.1 Failure Mode

Load at 1.75 mm varied between 4 and 15 kN for all beams tested. While the short-time load deformation curves for steel fibre reinforced beams are constantly decreasing with increasing de-flection, the curve for synthetic fibre reinforced beams may reach a maximum between 0.5 and 2 mm deflection, depending on the fibre type. But in the testing procedure a load at 1.75 mm is fixed and the sustained load is fixed by a percentage of this load. The results obtained from individual creep tests performed in these investigations are therefore different from those using 3 beams packed together and loaded with a mean value of the load.

Beams with synthetic fibres show big differences in the results, depending on fibre type. Actually, fibres type B (Fig. 9) showed very high deformations, but carrying the sustained load for long time, while type F (Fig. 6) fibres showed small deformations and up to now, only one sample reached its creep resistance after 2316 days of loading at 60 %. Most synthetic fibres reached the tertiary creep stage [1] after a quite long period and failed. The failure mode is not easily identified, as an investigation of the fibres with different sizes is not that easy after the test. However, if the elongation of the polymer gets too high, it can be expected that the fibre material will fail, maybe partly combined with a pull out of the fibres. Beams with the investigated steel fibres with end hooks, showed the lowest deformation in this post crack creep tests. However, there is also an upper load limit for these fibres. Increasing the load to 75 % through 85 % results in immediate creep rupture, due to fibre pull out. The pull out action can be heard during testing by a clicking noise.

5 Conclusions

For durability and serviceability reasons, design of FRC structures has to avoid too high long term deformations and creep rupture. Therefore more knowledge regarding creep of cracked FRC is necessary. Up to now no international specification for tests is available. Numerous tests according to the Austrian Guideline "Fibre reinforced Concrete" have been performed with different fibres, but same dosage. Unnotched beams with a span of 450 mm, precracked to 1.75 mm deflection were loaded with 50 % of short time residual load, which was stepwise increased to higher levels.

The results give a clear indication, how cracked FRC will behave under sustained load. However, the test results may only be applied to the fibre dosages and the quite high crack mouth opening displacement used in these tests prior to the creep test. Creep and shrinkage of the concrete was not eliminated from the data.

The used testing equipment stood the test. The supports have been improved recently. The influence of the storage climate has still not proved satisfactory.

Following above stated test conditions it may be concluded for the tested fibres: Beams with low quality synthetic fibres will not withstand a sustained load of 50 % of short time residual strength. Beams with most synthetic fibres will perform well at a load level of 50 %, but experience problems at a load level of 60 %. Creep rupture may occur immediately or after some years of loading. Therefore, it is difficult to predict the creep behaviour in advance. Beams with some synthetic fibres tend to suffer very high deformations not ending up in failure before some years of loading have passed. Beams with the hooked end steel wire fibres started with a slip out from the matrix and creep rupture occurred within several hours at load levels of 75–85 %.

Acknowledgments These tests were partly sponsored by members of the committee working with the guideline "Fibre Reinforced Concrete" of the Austrian Society for Construction Technology (obv): obv, Gueteverband Transportbeton, KrampeHarex Fibrin, Forta, Adfil, Grace, Bekaert, Cemex, Arcelor Bissen, Asamer & Hufnagel and Transportbeton.

References

1. Kusterle, W.: Creep of fibre reinforced concrete—flexural test on beams. In: Proceedings of Fibre Concrete 2015, CTU in Prague, Faculty of Civil Engineering, 10–11 Sept 2015, Prague
2. DAfStb: Guideline steel fibre reinfoced concrete (in German Richtlinie Stahlfaserbeton). Beuth-Verlag, Berlin (1996)
3. obv (Austrian Society for Construction Technology): Guideline Fibre Reinforced Concrete (in German: Richtlinie Faserbeton), Vienna (2008)
4. Bast, T., Eder, A.: Untersuchungen zum Langzeitstandverhalten von gerissenen Faserbetonen unter Biegezugbeanspruchung. Diploma-thesis, OTH Regensburg (2007)
5. Kusterle, W.: Viscous material behaviour of solids—creep of polymer fibre reinforced concrete. In: Proceedings of the 5th Central European Congress on Concrete Engineering. obv, Baden (2009)

Experiences from 14 Years of Creep Testing of Steel and Polymer Fiber Reinforced Concrete

S. Van Bergen, S. Pouillon and G. Vitt

Abstract This paper presents an overview of different creep setups, applied in the Bekaert concrete laboratory since 2002. The tests demonstrate the long term behaviour of cracked fibre reinforced concrete under sustained load, investigating statically determinate and indeterminate test setups, fibre type (steel and polymer macro fibres) and fibre dosage. The duration of the creep tests ranges from less than 1 to 14 years. The following observations were made: (1) Specimens reinforced with polymer macro fibres show a large deformation increase under sustained load, while specimens reinforced with steel fibres withstand. (2) After more than 2 years of testing, no difference in behaviour could be observed between steel fibres with different anchorage and corresponding material characteristics. (3) Increased deformation during the pre-cracking stage, along with increased sustained load has been found to have a higher impact for polymer macro fibres than steel fibres. All setups have been tailored to specific applications and thus vary in size, applied load and pre-cracking deformation. The same holds for setups used by other research groups so that it is difficult to compare results. Based on the experience built up during the course of the programs, proposals for an improved test setup are made.

Keywords FRC · Fibres · Creep · Steel · Polymer

1 Introduction

Creep is a term used to describe how a material deforms over time when subject to a sustained loading. Effects from creep could be increasing deformation with time, as well as a loss of load bearing capacity. Within the scope of this paper, the term

S. Van Bergen (✉) · S. Pouillon · G. Vitt
NV Bekaert SA, Kortrijk, Belgium
e-mail: sarah.vanbergen@bekaert.com

S. Pouillon
e-mail: steven.pouillon@bekaert.com

G. Vitt
e-mail: gerhard.vitt@bekaert.com

P. Serna et al. (eds.), *Creep Behaviour in Cracked Sections of Fibre Reinforced Concrete*, RILEM Bookseries 14,
DOI 10.1007/978-94-024-1001-3_4

creep refers to the increasing deformation of a pre-cracked fibre reinforced concrete (FRC) specimen due to a long term sustained load. Creep failure, in that context, refers to the bending failure of the pre-cracked FRC specimen due to the sustained load.

Steel fibre reinforced concrete (SFRC) has been successfully used as a construction material for decades without known issues related to long term effects such as creep. Cold drawn steel wire fibres with positive mechanical anchorage are the most relevant and governing fibre type in regular concrete. However, other fibre materials and shapes are being introduced which substantially differ from the typical steel fibres in both material properties and shape. The analysis of creep behaviour of FRC as a composite material can be split up into the creep properties of the fibres, the concrete and the interface and anchorage of the fibres in the concrete. Whereas steel itself does not exhibit relevant creep under normal service conditions (<370 °C, ≪ultimate tensile strength), polymers experience creep already at room temperatures, with the rate of creep varying significantly with polymer type, temperature and applied load. On the other hand, the creep of concrete under compressive load is attributed by Hilaire et al. [1] to short-term micro-diffusion of water between capillary pores, sliding of C-S-H sheets and micro-cracking and occurs at all stress levels. It is therefore subject to secondary effects such as pore water content and hardening of unhydrated cement. Because of the aforementioned reasons, the amount of fibres, the curing regime, concrete type, age of loading the specimen and relative humidity in the test environment are commonly brought forward as factors which could impact the rate of concrete creep in the specimen. Knowing the creep properties of polymer materials used to manufacture fibres for concrete, similar creep behaviour could be expected for the corresponding cracked polymer fibre reinforced concrete (PFRC). Additionally the impact of test boundary conditions can't be neglected in this case. Nevertheless, only little experience is available so far. Below, an overview is presented of the most relevant research programs related to the creep behaviour of pre-cracked FRC specimens, run in the Bekaert concrete laboratory (Zwevegem, Belgium) since 2002. In these programs statically determinate and indeterminate test setups are applied. The age of the test specimens ranges from less than 1 to 14 years.

2 Materials

Polymer fibres in this paper are anonymized and identified by their length (L) and aspect ratio (L/d), while steel fibres are additionally identified by their anchor type and wire characteristic (Table 1). The 5D fibre distinguishes itself from other fibres by its double hooked end, high tensile strength and additional higher ultimate strain of 6 %.

Three concrete mixes are applied (Table 2); varying in compressive strength, aggregate type and maximum aggregate size. Specimens were cured for at least 28 days under water before the pre-cracking stage to allow for better repetitiveness of results. However panels in programs B2 and B3 were pre-cracked and set up

Table 1 Fibre characteristics

Fiber ID	Material	L	L/d	f_u	E	ρ	Anchor type
		(mm)	(–)	(MPa)	(GPa)	(kg/m^3)	
PF–83/50	Polymer	50	83	620	9.5	920	Embossed
PF–90/40	Polymer	40	90	620	9.5	920	Embossed
PF–48/48	Polymer	48	48	520	10	920	Embossed
PF–73/65	Polymer	65	73	640	10	920	Embossed
PF–80/54	Polymer	54	80	580	4.7	920	Embossed
3D 45/30BG	Steel	30	45	1270	200	7800	3D anchor
3D 65/35BG	Steel	35	65	1345	200	7800	3D anchor
3D 65/60BG	Steel	60	65	1160	200	7800	3D anchor
3D 80/60BG	Steel	60	80	1225	200	7800	3D anchor
5D 65/60BG	Steel	60	65	2300	200	7800	5D anchor

Table 2 Overview of different concrete compositions (dry weights, kg/m^3)

Concrete	C1	C2	C3
	C35/45	C30/37	C25/30
CEM I 42,5 R HES	427	350	280
River sand 0/4	854	822	822
Broken limestone 4/7	854	456	–
Round gravel 5/15	–	–	1003
Broken limestone 7/14	–	547	–
Water	214	186	168

after only 2 days to simulate conditions related to shotcreting. All specimens are exposed to minimum room temperature of 16 °C. There is no control of air temperature or relative humidity.

3 Test Programs

3.1 *Statically Determinate Beam Test*

The purpose of the statically determinate beam test and corresponding creep setup is to obtain a comparison of the mechanical properties of different cracked FRC materials under long term sustained loading. As described in Table 3, all specimens are pre-cracked up to a predefined deflection (δ_{pc}) or crack mouth opening displacement ($CMOD_{pc}$) in either a 3-point or 4-point deflection controlled bending test (Fig. 1) to define their residual flexural load (F_r). The specimens are then moved to a permanent setup where they are subjected to their sustained load (F_{creep}), which is defined as a fraction of F_r at δ_{pc} or $CMOD_{pc}$. During the creep phase, the additional deflection due to creep (δ_{creep}) of all specimens is measured at regular time intervals.

Table 3 Overview of statically determinate beam tests

Creep program	A1	A2	A3
Start program	May 2002	July 2010	October 2012
End program	August 2015	*Ongoing*	June 2015
Object size	L = 500 mm	L = 870 mm	L = 600 mm
	W = 150 mm	W = 1500 mm	W = 150 mm
	H = 50 mm	H = 40 mm	H = 150 mm
Type of test	4-point bending	3-point bending	3-point notched EN14651
F_{creep}	50 % F_r	60 % F_r	60 % F_r
δ_{pc} or $CMOD_{pc}$	δ_{pc} = 5 mm	δ_{pc} = 3 mm	$CMOD_{pc}$ = 0.5 mm
# specimens	9	3	6
# failed during test	5 PFRC	1 PFRC	–
Concrete type	C1	C2	C2
Curing days	42 days	42 days	42 days
Fiber types	3D 65/35BG	3D 45/30BG	3D 65/60BG
	PF–48/48	PF–90/40	5D 65/60BG
	PF–83/50	PF–80/54	

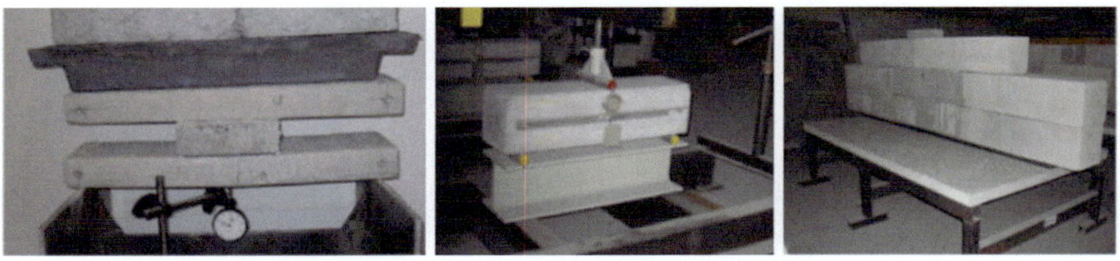

Fig. 1 *From left to right* setup program A1; program A3; program A2

As shown in Table 5, programs A1 and A3 contain a series of 3 specimens per fibre type. However, specimens which were accidentally overloaded in the pre-cracking phase were excluded from the test program. The average deflection due to creep ($\delta_{creep,m}$) presented in this report is thus the average of the long term deflections of specimens included per series at the moment of failure or last measured point. Since all programs are executed with different setup and specimen size, the actual measured deflections out of different programs can't be compared as such.

The first two test programs focus on the comparison between PFRC and SFRC, while the third program aims at a more standardised setup in line with the notched 3 point bending test EN14651 [2], where the influence 3D versus 5D fibres on the creep behaviour is investigated.

3.2 *Statically Indeterminate Plate Test*

In addition to the statically determinate beam tests, the long term behaviour of SFRC and PFRC plates on continuous support has been investigated (Table 4). The focus of this investigation is directed to the analysis of the long term behaviour of tunnel linings under rock pressure around an anchor bolt. For this purpose, a long term test setup is built so as to load cast panels according to EN14488-5 [3]. Per tested series, 3 FRC square plates of 600 mm × 600 mm × 100 mm are centrally loaded in a displacement controlled manner as shown in Fig. 2. The pre-cracking test is ended at a specified δ_{pc} (Table 4) of the centre point of the panel. The maximum load and thus the relevant crack pattern are usually achieved at deflections from 2 to 3 mm. But while in some cases 2 mm deflection is sufficient to

Table 4 Overview of centrally loaded plate tests

Creep program	B1	B2	B3
Start program	September 2009	March 2012	April 2012
End program	September 2012	*Ongoing*	January 2013
Object size	W = 600 mm	W = 600 mm	W = 600 mm
	H = 100 mm	H = 100 mm	H = 100 mm
Type of test	EN14488-5	EN14488-5	EN14488-5
F_{creep}	60 % F_r	40–60–80 % F_r	60 % F_r
δ_{pc}	3 mm	2 mm	3 mm
# specimens	12	6	6
# failed during test	3 PFRC	–	–
Concrete type	C3	C1	C1
Curing days	58–62 days	2 days	2 days
Fiber types	3D 65/35BG	3D 65/35BG	3D 65/35BG
	3D 80/60BG	PF–73/65	PF–73/65
	PF–90/40		

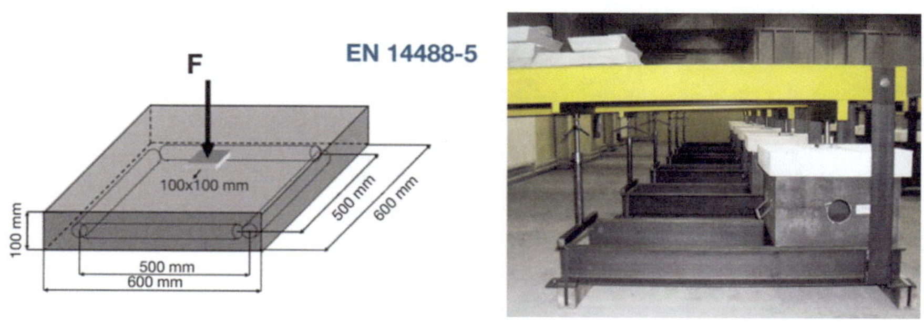

Fig. 2 Setup of EN14488-5 creep test method

reach this level, a deflection of 3 mm is preferred. Otherwise the specimen could still be in the phase of initial cracking, not having developed the final crack pattern.

The load-deflection curve is recorded and F_r is registered. The deflection recovery was not measured. The plate is then moved to the permanent setup of a rigid steel square loading block having a contact outline of (100 ± 1) mm × (100 ± 1) mm and where F_{creep} is applied, equivalent to 40–60 % of F_r. F_{creep} in program B2 was adjusted over time up to 80 % F_r as shown in Fig. 7. The deflection is measured at regular intervals. $\delta_{creep,m}$ (Table 6) is, similarly to the beam tests, the average of the individually registered deflections due to creep at failure or last measured point. The fibres which are taken up in programs B2 and B3 are fibres commonly used in shotcrete applications. Specimens with 25 kg/m^3 3D 65/35BG and 7 kg/m^3 PF-73/65, were loaded with similar F_{creep} as shown in Table 7.

Table 5 Overview of statically determinate beam tests

Program	Fiber type	Specimens	Dosage		Average test duration	$\delta_{creep,m}$
			kg/m^3	% vol	days	$\times 10^{-2}$ mm
A1	PF–48/48	3	4.55	0.49	1694	841
	PF–83/50	3	4.55	0.49	764	947
	3D 65/35BG	2	20	0.26	4318	195
A2	3D 45/30BG	1	30	0.38	1967	191
	PF–80/54	1	6	0.65	674	874
	PF–90/40	1	6	0.65	1966	401
A3	3D 65/60BG	3	40	0.51	719	28
	5D 65/60BG	2	40	0.51	738	30

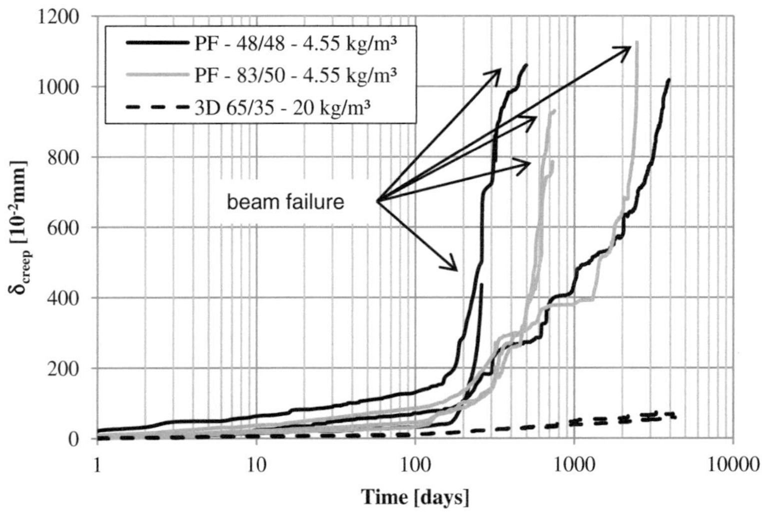

Fig. 3 Program A1: 4-point bending on 150 mm × 50 mm × 500 mm at 50 % F_r (δ_{pc} = 5 mm)

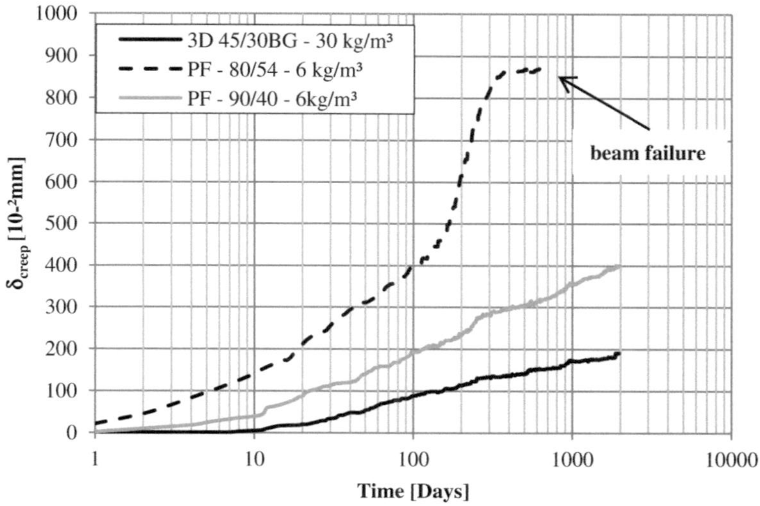

Fig. 4 Program A2: 3-point bending on 1500 mm × 870 mm × 40 mm at 60 % F_r (δ_{pc} = 3 mm)

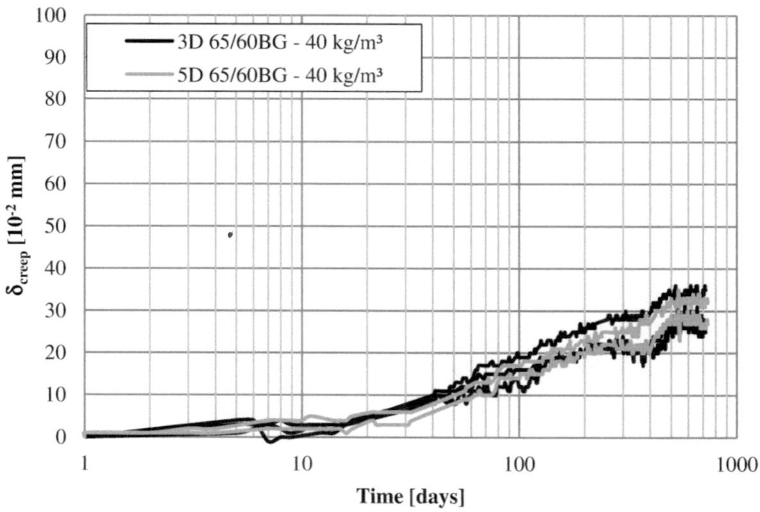

Fig. 5 Program A: 3-point bending test (based on EN14651) at 60 % F_r ($CMOD_{pc}$ = 0.5 mm)

Table 6 Overview of statically indeterminate plate tests

Program	Fiber type	Specimens	Dosage		Average test duration	$\delta_{creep,m}$
			kg/m^3	% vol	days	mm × 10^{-2}
B1	3D 80/60BG	3	25	0.32	890	119
	PF–90/40	3	4.5	0.49	301	569
	3D 80/60BG	3	40	0.51	880	166
	3D 65/35BG	3	40	0.51	885	158
B2	3D 65/35BG	3	25	0.32	1392	111
	PF–73/65	3	7	0.76	1385	274
B3	3D 65/35BG	3	25	0.32	281	43
	PF–73/65	3	7	0.76	273	81

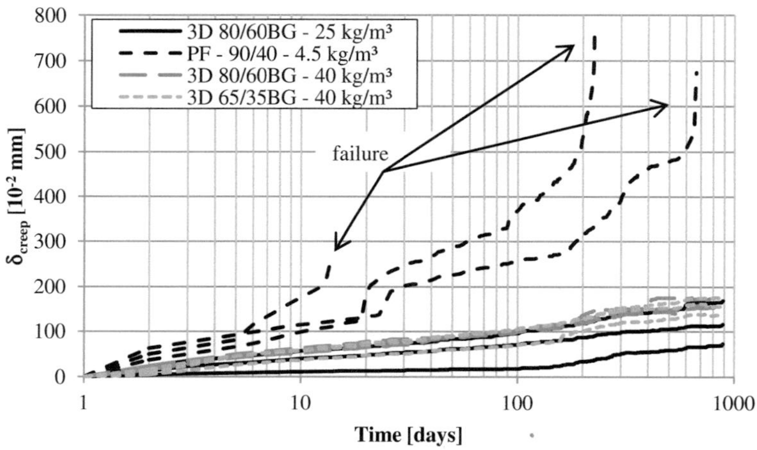

Fig. 6 Program B1: EN14488-5 on 600 mm × 600 mm × 100 mm at 60 % F_r (δ_{pc} = 3 mm)

4 Program Results and Discussion

4.1 Statically Determinate Beam Test

As demonstrated in Table 5, $\delta_{creep,m}$ in the programs A1 and A2 of the PFRC specimens is 2–4 times that of the SFRC specimens. In program A1, all PFRC specimens show an exponential deformation increase and 5 out of 6 specimens eventually fail (Fig. 3). However, the rate of creep is not so easy to predict as can be concluded from program A2 (Fig. 4) where both tested PFRC with identical dosage show very different results. It should therefore be concluded that big variations can be found among polymer fibres. These differences cannot be explained by their material properties alone; shape and surface geometry might also play a role.

According to the results found in program A3 (Fig. 5), one could conclude that the 3D fibres and the 5D fibres show the same behaviour. However, as the wire strength in a 5D fibre is much higher and also much more utilized, the 5D anchor and the concrete compression zone has to withstand a higher actual load. With an increasing F_{creep}, deflections are expected to be higher. However an overall rise in $\delta_{creep,m}$ could not observed.

In all programs, fluctuations can be observed in the registered δ_{creep}. These fluctuations are in this case attributed to variations in temperature and relative humidity in the test room.

4.2 *Statically Indeterminate Plate Test*

The centrally loaded plate tests confirm the observations of the flexural tests, as PFRC specimens consistently show higher $\delta_{creep,m}$ up to 4 times higher than SFRC specimens in identical test programs (Table 6). In test program B1, all PFRC specimens failed as well and the actual time to failure shows a large scatter with failure occurring at 13, 225 and 666 days (Fig. 6). The 3D 80/60BG steel fibre was tested in the same setup at dosage rates of 25 and 40 kg/m^3. With this increasing amount of steel fibres, the scatter of δ_{creep} results reduces from 40 % to about 15 % while the $\delta_{creep,m}$ slightly increases.

The latter could be an effect from creep in the compression zone, because the absolute F_{creep} is higher for the higher fibre dosages. Thus deformations originating from creep in the compression zone are expected to be higher. When comparing the

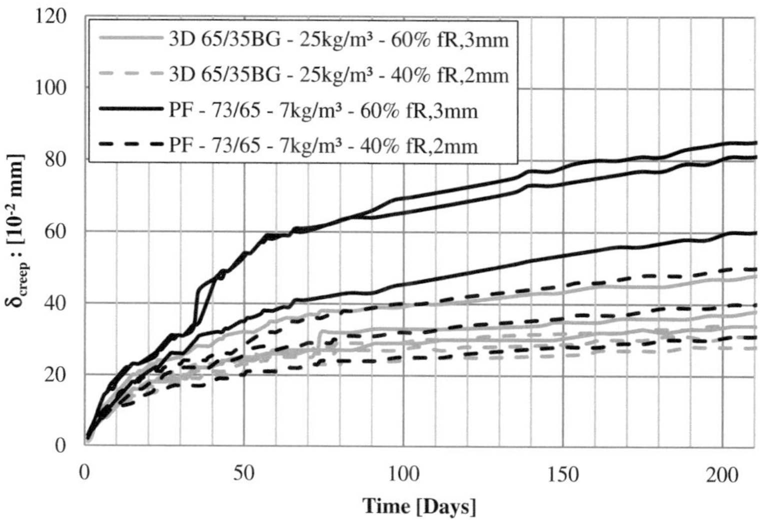

Fig. 7 EN14488-5: Effect of load level and pre-cracking deflection on δ_{creep} (B2; B3)

Table 7 $F_{creep,m}$ and $\delta_{creep,m}$ at 250 days (B2; B3)

Fiber type	Dosage	$F_{creep,m}$		$\delta_{creep,m}$
	kg/m^3	–	kN	mm × 10^{-2}
3D 65/35BG	25	40 % F_r δ_{pc} = 2 mm	20	33
PF–73/65	7		18	43
3D 65/35BG	25	60 % F_r δ_{pc} = 3 mm	32	42
PF–73/65	7		30	81

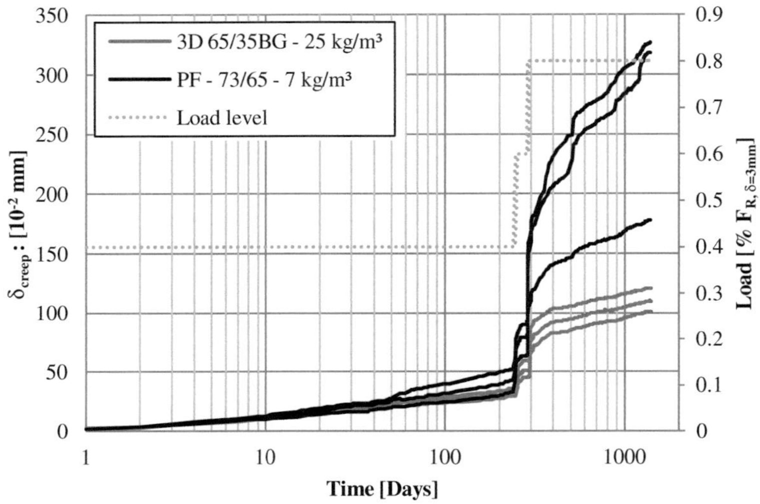

Fig. 8 Program B2: EN14488-5 on 600 mm × 600 mm × 100 mm at 40–60–80 % F_r (δ_{pc} = 2 mm)

results of 40 kg/m^3 3D 80/60BG and 3D 65/35BG, δ_{creep} and its variation are of the same order. When comparing programs B2 and B3 (Table 7), there is an actual increase of the average sustained load per series ($F_{creep,m}$) by 50 % while increasing the δ_{pc} from 2 to 3 mm.

As can be seen in Fig. 7, the effect is more important in the case of PFRC, with an increase of 90 % compared to 30 % for SFRC. Test program B2 shows similar results, where the impact of the load intensity on the creep rate seems to be of major importance for PFRC, but less for SFRC. Program B2 is, as mentioned before, still running. In this program, as shown in Fig. 8 the observed δ_{creep} of PFRC at 40 % of F_r δ_{pc} = 2 mm was shown to be 30 % higher than that of SFRC specimens. However once the load was increased to 60 % F_r δ_{pc} = 2 mm and higher, the rate of creep of the specimens with 7 kg/m^3 PF—73/65 quickly rose to more than two times of that of the SFRC specimens. The influence of the pre-cracking and testing at early age can't be compared due to the fact that different fibres and concrete were applied in the older setup B1.

5 Conclusions

Since the above described tests were one by one tailored to a specific applications, individual test results from different programs can't be compared as such. More in depth analysis needs to be done to better understand creep in FRC materials, but before extra tests are performed, it is essential to agree on a single test setup that can give the material characteristics required for creep analysis in structural design. However, a number of important conclusions can be drawn from the research described above and some of them are helpful for the definition of a unified and improved test setup:

- Depending on the specimen type and test setup, the tested PFRC specimens show up to 4 times higher $\delta_{creep,m}$ than the tested SFRC specimens.
- The majority of tested PFRC specimens fail under the applied sustained loading while none of the SFRC specimens collapse.
- SFRC with 3D and 5D steel fibres show similar creep behaviour.
- The time to failure of PFRC specimens cannot be predicted from the test data since it is varying a lot. Thus the tests need to be performed during a sufficiently long time range and need to include fibre characteristics such as shape and surface geometry.
- There are indications that higher fibre dosages result in additional deflection. This is most likely due to the higher absolute F_r which is achieved at higher fibre dosages.
- The initial crack width and F_{creep} have an important effect on the creep behaviour.
- The combined effect of increased δ_{pc} and increased F_{creep} has been found to have a bigger impact for PFRC than for SFRC.
- Due to the indeterminate crack pattern in the plates, which is inherent to the fully supported setup, it is difficult to interpret the actual creep behaviour of the material, rather than that of the system. The initial test deflection and (final) crack pattern strongly affects the performance in the long term test setup. Additionally, external factors such as friction of test supports and arching is likely to have an influence. Therefore, interpreting results of these centrally loaded plates is a lot more complex than that of a notched beam according to EN14651. The indeterminate plate test appears less suitable for the comparisons of FRC material properties.
- The statically determinate beam tests with a clear load path, however, give an easily interpretable result. Especially the notched 3 point bending beam test based on EN14651 give allow for reproducible test results and identification of material properties. In addition, the underlying test method is already well known in the European industry (EN 14651).

- A potential effect of F_{creep} on the short-term ultimate performance has not been investigated in the described tests.
- The effect of increased temperature levels has not been investigated in the described tests. In view of the thermoplastic material properties of most polymer macro fibres, this parameter should also be important and needs to be considered in an improved test setup.

References

1. Hilaire, A., Benboudjema, F., Darquennes, A., Berthaud, Y., Nahas, G.: Analysis of concrete creep in compression, tension and bending: numerical modelling. Ninth International Conference on Creep, Shrinkage, and Durability Mechanics (CONCREEP-9)
2. EN14651:2005: Test method for metallic fibred concrete. Measuring the flexural tensile strength (limit of proportionality (LOP), residual), 11 July 2005
3. EN14488-5: Testing sprayed concrete—Part 5: determination of energy absorption capacity of fibre reinforced slab specimens, 27 Feb 2006

Creep Deformations of Structural Polymeric Macrofibers

Rutger Vrijdaghs, Marco di Prisco and Lucie Vandewalle

Abstract Fiber reinforced concrete (FRC) is a cementitious composite in which fibers are added to the fresh concrete to improve its behavior. Fibers can be added to bridge crack faces, thus increasing the residual load-bearing capacity of a concrete element. Furthermore, these fibers can arrest crack growth and increase the long-term durability of the structural element. However, long-term durability can be compromised by time-dependent phenomena such as creep. In FRC, time-dependent crack widening can be mainly attributed to two mechanisms: fiber creep and gradual fiber pull-out from the concrete matrix. In the case of polymeric fibers, fiber creep in the crack may contribute significantly to the crack widening. This paper presents the experimental results of creep tests on two different commercially available polypropylene macrofibers. Different sustained load levels are considered, ranging from 22 to 63 % of the fiber strength. The results show that sudden failure occurs in the secondary creep phase at all load levels. Furthermore, the time to failure and the total strain at failure depend very strongly on the applied load level. The total creep strain at failure may become very large: creep coefficients greater than 10 have been observed, especially at lower load levels.

Keywords Fiber reinforcement · Polymeric fiber · Creep · Experimental

1 Introduction

Fiber reinforced concrete (FRC) is a composite material in which fibers are added to the fresh concrete mix [1, 2]. These fibers can improve the properties of the concrete in the fresh or hardened state. In structural applications, fibers can partially or totally replace the traditional reinforcement. For these purposes, the fibers provide

R. Vrijdaghs (✉) · L. Vandewalle
Department of Civil Engineering, KU Leuven, Leuven, Belgium
e-mail: rutger.vrijdaghs@kuleuven.be

M. di Prisco
Department of Structural Engineering, Politecnico di Milano, Milan, Italy

P. Serna et al. (eds.), *Creep Behaviour in Cracked Sections of Fibre Reinforced Concrete*, RILEM Bookseries 14,
DOI 10.1007/978-94-024-1001-3_5

an enhanced post-cracking tensile strength in the hardened state by bridging crack faces [3]. Commercially available fibers can be made from a number of different materials: steel, glass, synthetic and natural being the most common types [4]. Steel fibers are widely used in structural applications [5–8]. However, steel fiber reinforced concrete (SFRC) may not be suitable for all applications, and synthetic fiber reinforced concrete (SyFRC) [7] and glass FRC (GFRC) [9] have been successfully used as well.

While considerable research effort has been devoted to the material parameters and short-term structural characteristics of FRC [10–16], the long-term behavior of FRC elements is still poorly understood [17–19]. One key element in long-term performance is creep, as time-dependent crack growth and crack widening may decrease durability. In particular, in cracked FRC elements, the fibers bridging crack faces are transmitting forces and as such, they are subjected to possible long-term loading. In the case of polymeric fibers, these sustained loads can give rise to creep deformations of individual fibers. In an FRC element, the fiber's own deformation can influence the crack width growth and therefore, it should be taken into account.

2 Experimental Setup

2.1 Material Characterization

In this research, two different polypropylene fibers are considered, called type A and type B in this paper. Type B is a generational upgrade to type A with improved tensile strength. These fibers are certified to be used in structural applications according to the European Standard EN 14889-2 [20]. Both types are commercially available in Belgium. The length of the commercial fiber is 45 mm in both cases. The equivalent diameter is 0.95 and 0.9 mm for type A and B, respectively. Displacement controlled tensile tests were done to characterize the material, i.e. to determine the σ–ε curve. The test setup is shown in Fig. 1a. Uncut fibers, (3) in Fig. 1a, are clamped in a clamping setup, (2) in Fig. 1b, that uses a combination of gluing and clamping to minimize fiber end slippage. The uncut fibers were obtained from the manufacturer before cutting to commercially available lengths (45 mm). The displacement is measured by the internal gauge of the test machine, and an external load cell (capacity: 900 N, accuracy: ±0.05 %) is used to record the applied force, see (1) in Fig. 1b.

The tests characteristics are shown in Table 1. Fiber type A has been tested according to the European standard EN 14889-2 [20], while type B has been tested in accordance with the (yet unpublished) revision proposal of that standard, which imposes a higher testing rate of 50 %/min. The data acquisition rate was 10 Hz for both fiber types.

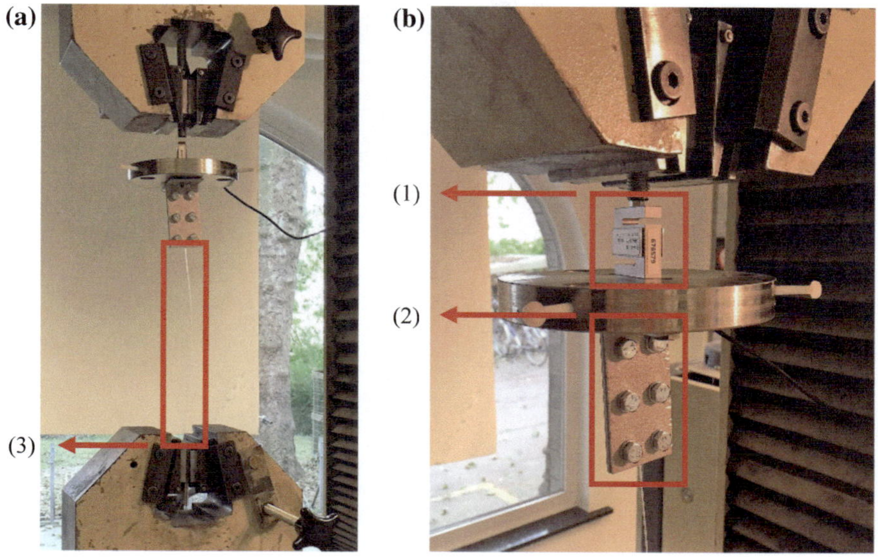

Fig. 1 **a** Overview of the characterization setup **b** close-up of the load cell and clamping system with (*1*) load cell (*2*) clamping system (*3*) fiber

Table 1 Test characteristics for both fibers

	Testing speed (mm/min)	Initial fiber length (mm)	Number of samples
Type A	10	430	6
Type B	125	250	5

2.2 *Creep Frame*

After the characterization of the fibers, both types are subjected to creep tests in a climate chamber at constant temperature (20 °C) and relative humidity (60 %). Uncut fibers were clamped on both sides using the same clamping system as used in the characterization tests. The fiber is then placed in a creep frame, and subjected to a sustained load until failure. The time-dependent fiber elongation is contactlessly measured by a laser sensor at a rate of 1 Hz in the first week, and at 0.1 Hz afterwards. The maximum measurable elongation is 125 mm. The creep setup is schematically shown in Fig. 2a and a photo of the actual frame is presented in Fig. 2b.

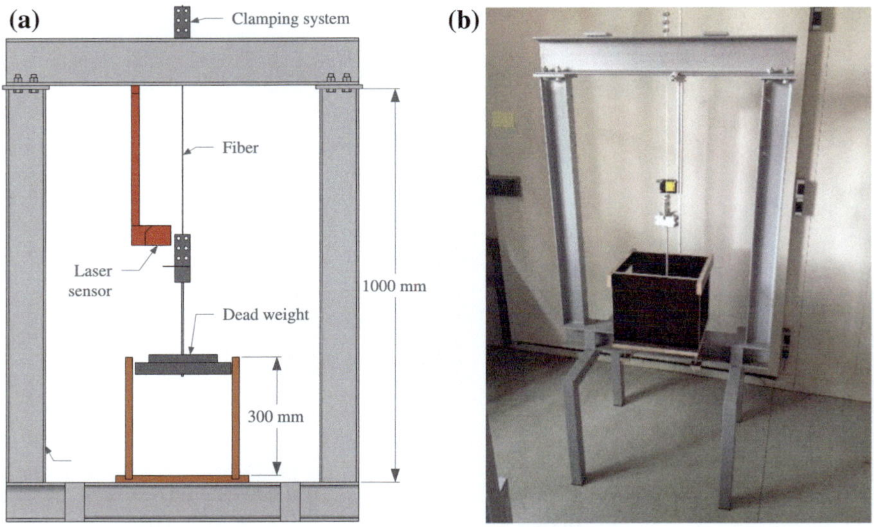

Fig. 2 **a** Simplified creep frame, not shown is the supporting base **b** photo of the creep frame

2.3 *Creep Specimens*

In total, 26 creep samples have been tested, 14 of fiber A and 12 of fiber B. Every sample had an initial length of 200 mm and 5 different load levels were considered: 22, 36, 43, 53 and 63 %. The load level is defined as the applied creep load with respect to the average tensile strength. The number of specimens per load level and fiber type are summarized in Table 2.

3 Results

3.1 *Material Characterization*

The σ–ε curves for type A and B are shown in Fig. 3a, b, respectively. Furthermore, the fiber properties can be found in Table 3. The Young's modulus is

Table 2 Number of creep specimens per load level and fiber type

Load level (%)	Type A	Type B
22	0	2
36	4	2
43	4	2
53	6	2
63	0	4
Total	14	12

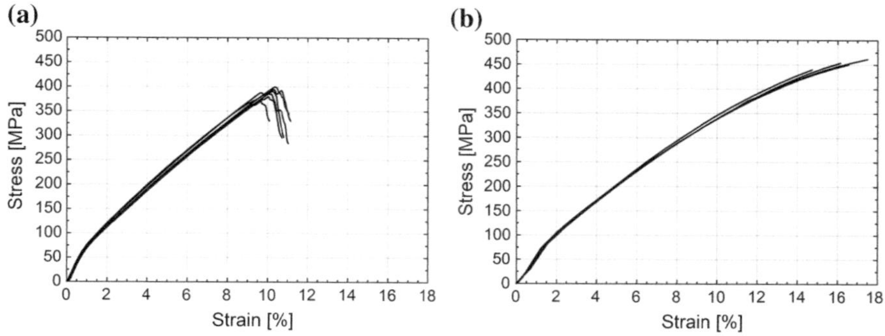

Fig. 3 σ–ε curve for **a** type A and **b** type B

Table 3 Mechanical properties of both fibers

	Type A	Type B
Average strength (CoV[a])	387 MPa (2.8 %)	451 MPa (1.8 %)
Average strain at maximum stress (CoV)	10 % (2.6 %)	16 % (6.2 %)
Average Young's modulus (CoV)	6200 MPa (2.7 %)	4400 MPa (2.3 %)

[a]CoV: coefficient of variation, ratio of the standard deviation w.r.t. the mean

calculated according to the revision proposal of EN 14899-2. It is noted that the strength and Young's modulus is relatively low as compared to other commercially available macro-synthetic fibers.

3.2 Creep Deformations

The creep results of fiber type A and B are shown on a logarithmic time scale in Fig. 4a, b, respectively. In these figures × and ■ denote fiber creep failure, the former representing failure in the middle of the fiber, the latter failure near the clamps. O denotes an aborted test due to time constraints. Note that for both fiber types, the strain axis is the same to highlight differences more clearly.

The average time to failure and strain at failure, $t_{failure}$ and $\varepsilon_{failure}$ respectively, are summarized in Table 4.

It is clear that the load level significantly influences the time to failure as well as the total strain at failure. Overall, very large strains can be expected in relatively short periods of time: the time to failure ranges from several hours to several months for fibers loaded at 63 or 43 % of their strength.

In all cases however, failure is sudden with no noticeable strain acceleration on a linear time scale (contrary to the logarithmic time scale in Fig. 4), that is usually associated with tertiary creep.

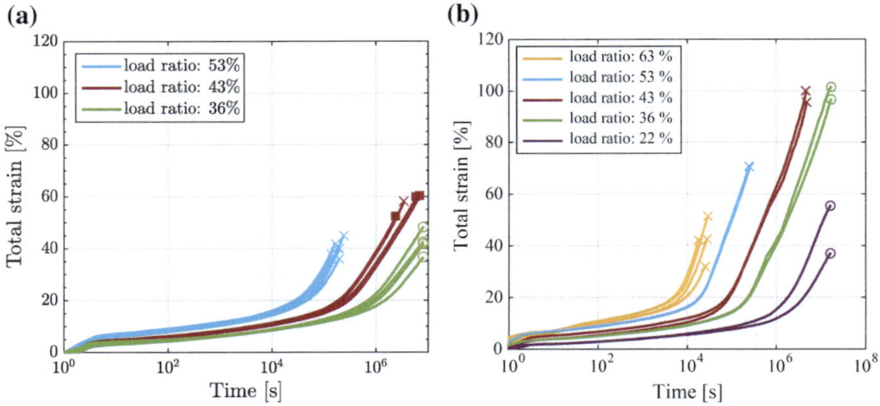

Fig. 4 Creep results of fiber type **a** A and **b** B

Table 4 Creep results of fiber

Load ratio (%)	$t_{failure}$ (CoV)	$\varepsilon_{failure}$ (CoV)
Type A		
36	>8.1 × 10^6 s (–)	>43 % (11 %)
43	4.1 × 10^6 s (49 %)	58 % (6.3 %)
53	1.9 × 10^5 s (16 %)	40 % (7.8 %)
Type B		
22	>1.6 × 10^7 s (–)[a]	>46 % (–)[a]
36	>1.6 × 10^7 s (–)[a]	>99 % (–)[a]
43	4.6 × 10^6 s (–)[a]	98 % (–)[a]
53	2.5 × 10^5 s (–)[a]	70 % (–)[a]
63	2.4 × 10^4 s (20 %)	42 % (19 %)

[a]CoV cannot be calculated for 2 specimens

4 Discussion

All fibers exhibited very large creep deformations. The relative creep deformation can be expressed in terms of a creep coefficient φ_{creep}. This is the ratio of the creep strain divided by the instantaneous strain, see Eq. (1).

Table 5 Average creep coefficient for all experiments (shown in parentheses is the CoV)

Load level (%)	Type A	Type B
22	N/A	>14 (−)
36	>15 (12 %)	>15 (−)
43	16 (6.7 %)	12 (−)
53	8.8 (8.7 %)	7.7 (−)
63	N/A	4.4 (23 %)

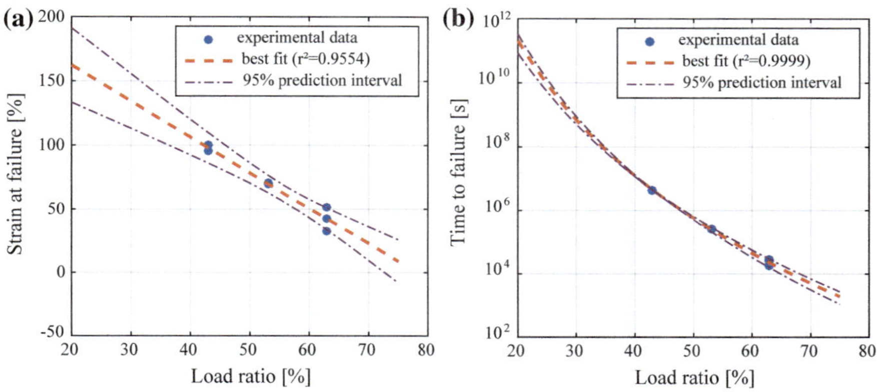

Fig. 5 Fiber type B extrapolation of the **a** strain at failure and **b** time to failure

Table 6 Expected strain at failure and time to failure for the unfinished experiments

Load ratio (%)	Expected strain at failure (95 % confidence bounds)	Expected time to failure (95 % confidence bounds)
22	157 % (129 %, 184 %)	5.4×10^{10} s (2.6×10^{10} s, 8.3×10^{10} s)
36	118 % (101 %, 134 %)	5.5×10^{7} s (4.7×10^{7} s, 6.3×10^{7} s)

$$\varphi_{creep} = \frac{\varepsilon_{creep}}{\varepsilon_{instant}} \tag{1}$$

The average creep coefficient φ_{creep} is shown in Table 5. In general, large creep coefficients are observed, and these coefficients are strongly dependent on the load ratio with a decrease in load ratio corresponding to an increase in creep coefficient.

With the results of type B, an estimation of the time to failure as well as the total strain at failure can be given for the fibers loaded at 36 and 22 %. The extrapolation of both parameters is shown in Fig. 5a, b. Using these models, and assuming extrapolation is valid, it is possible to calculate the expected time to failure as well as the corresponding total strain at failure for the aborted tests, i.e. at a load ratio of 22 and 36 %. The results are presented in Table 6.

There is good agreement between the fitted linear and power laws and the experimental data. These extrapolations underline the large expected creep strains. Furthermore, the results highlight the fact that the fiber's own creep strain can significantly widen a crack in an FRC element.

5 Conclusions

- Two types of commercially available structural polypropylene macrofibers have been subjected to sustained loading and the time-dependent deformations are recorded. All specimens had an initial length of 200 mm, and 5 different load levels were considered: 22, 36, 43, 53 and 63 % of the average strength of the fiber.
- The strength of the fiber was determined according to the European standard EN 14889-2 for both fiber types in a displacement controlled test.
- In total, 26 creep specimens were tested. The results have shown that the load ratio strongly influences the time to failure as well as the total strain at failure. For one fiber type, an extrapolation with respect to these two parameters has been done in order to estimate the failure characteristics.
- No tertiary creep, i.e. strain acceleration on a linear time scale, has been observed during the tests.
- The creep coefficient can be exceedingly large, and depends strongly on the load ratio.
- The results underline that the own fiber creep deformation can play an important role in the total crack width growth in FRC elements.

Acknowledgments The author would like to thank the financial support of the Agency for Innovation by Science and Technology in Flanders (IWT).

References

1. Balaguru, P., Shah, S.P.: Fiber-Reinforced Cement Composites. McGraw-Hill, Texas, USA (1992)
2. Bentur, A., Mindess, S.: Fibre Reinforced Cementitious Composites, vol. 449. Elsevier Science Publishers Ltd., England (1990)
3. di Prisco, M., Plizzari, G., Vandewalle, L.: Fibre reinforced concrete: new design perspectives. Mater. Struct. **42**(9), 1261–1281 (2009)
4. ACI Committee 544. In: Daniel, J.I. (eds.) State-of-the-Art Report on Fiber Reinforced Concrete. American Concrete Institute (2002)
5. Serna, P., et al.: Structural cast-in-place SFRC: technology, control criteria and recent applications in spain. Mater. Struct. **42**(9), 1233–1246 (2009)
6. Caratelli, A., et al.: Structural behaviour of precast tunnel segments in fiber reinforced concrete. Tunn. Undergr. Space Technol. **26**(2), 284–291 (2011)

7. Bernard, E.S.: Design of fibre reinforced shotcrete linings with macro-synthetic fibres. In: Shotcrete for Underground Support XI (2009)
8. de la Fuente, A., et al.: Experiences in Barcelona with the use of fibres in segmental linings. Tunn. Undergr. Space Technol. **27**(1), 60–71 (2012)
9. Ferreira, J.P.J.G., Branco, F.A.B.: The use of glass fiber-reinforced concrete as a structural material. Exp. Tech. **31**(3), 64–73 (2007)
10. *Fibre Reinforced Concrete*. In: Rossi, P., Chanvillard, G. (eds.) RILEM Proceedings of the 5th RILEM Symposium (BEFIB 2000), PRO15, BEFIB 2000. RILEM Publications S.A.R.L., Bagneux, France
11. *Fibre-Reinforced Concrete*. In: di Prisco, M., Felicetti, R., Plizzari G.A. (eds.) RILEM Proceedings of the 6th RILEM Symposium (BEFIB 2004), PRO39, BEFIB 2004. RILEM Publications S.A.R.L., Bagneux, France
12. *Fibre Reinforced Concrete: design and applications*. In: Gettu R. (ed.) BEFIB 2008. Rilem Publication S.A.R.L., Bagneux, France
13. Grzybowski, M., Shah, S.P.: Shrinkage cracking of fiber reinforced concrete. ACI Mater. J. **87** (2), 138–148 (1990)
14. Pasini, F., et al.: Experimental study of the properties of flowable fiber reinforced concretes. In: di Prisco, M., Felicetti, R., Plizzari, G.A. (eds.) 6th International RILEM Symposium on Fibre Reinforced Concretes, pp. 279–288. RILEM Publications S.A.R.L. (2004)
15. Barros, J., et al.: Post-cracking behaviour of steel fibre reinforced concrete. Mater. Struct. **38** (1), 47–56 (2005)
16. Buratti, N., Mazzotti, C., Savoia, M.: Post-cracking behaviour of steel and macro-synthetic fibre-reinforced concretes. Constr. Build. Mater. **25**(5), 2713–2722 (2011)
17. Babafemi, A.J., Boshoff, W.P.: Time-dependent behaviour of pre-cracked polypropylene fibre reinforced concrete (PFRC) under sustained loading. Res. Appl. Struct. Eng. Mech. Comput. 1593–1598 (2013)
18. Babafemi, A.J., Boshoff, W.P.: Tensile creep of macro-synthetic fibre reinforced concrete (MSFRC) under uni-axial tensile loading. Cement Concr. Compos. **55**, 62–69 (2015)
19. Zhao, G., di Prisco, M., Vandewalle, L.: Experimental investigation on uniaxial tensile creep behavior of cracked steel fiber reinforced concrete. Mater. Struct. 1–13 (2014)
20. CEN, EN 14889-2: Fibres for concrete—part 2: polymer fibres—definitions, specifications and conformity. European Committee for Standardization (2006)

Tensile Creep of Cracked Steel Fibre Reinforced Concrete: Mechanisms on the Single Fibre and at the Macro Level

W.P. Boshoff and P.D. Nieuwoudt

Abstract This paper reports on tests done to investigate the mechanisms causing the increased tensile creep of cracked fibre reinforced concrete (FRC) members when a sustained load is applied. It is important to understand the mechanisms as this will pave the way for improvements that can be made to reduce this creep and will also assist in creating prediction models as it will be based on the fundamental mechanisms involved. This paper presents results of uni-axial tensile creep tests of cracked steel fibre reinforced concrete (SFRC) and also tests at the single fibre level. Single fibres were embedded in the matrix and pull-out at different rates and sustained loading was also applied to the single, embedded fibres. It was found that the single fibre pull-out creep test results can be directly linked to the uni-axial tensile tests. It was also shown that the pull-out creep is proportional to the load up to at least 50 % of the ultimate load after which the pull-out creep increases non-linearly. This can be ascribed to micro-cracking around the hooked-end of the fibre. Lastly, it is postulated that the interfacial transition zone between the fibre hooked-end and the concrete matrix plays a significant role in the pull-out creep behaviour.

Keywords Fibre reinforced concrete · Creep · Creep mechanism · Micro-cracking · Single fibre tests

1 Introduction

Creep of a material can be defined as the variation of deformation of the material under sustained loading. This is a well-known phenomenon of concrete and has been researched for more than 80 years [1]. Some early publications highlighted the effect of environmental conditions [2] and many complex models exist today to predict the creep behaviour of concrete (e.g. [3]). Even though creep is not yet fully

W.P. Boshoff (✉) · P.D. Nieuwoudt
Department of Civil Engineering, Stellenbosch University, Stellenbosch, South Africa
e-mail: bboshoff@sun.ac.za

P. Serna et al. (eds.), *Creep Behaviour in Cracked Sections of Fibre Reinforced Concrete*, RILEM Bookseries 14,
DOI 10.1007/978-94-024-1001-3_6

understood, simplified models are used on a daily basis by structural engineers when designing concrete structures, e.g. [4].

Fibres have been researched for a number of decades as possible (partial) replacement of conventional reinforcement in concrete structures [5]. The *fib* Model Code 2010 [6] now includes design models for the use of fibre reinforced concrete (SFRC). A constitutive law for FRC in tension can be deduced from a relatively simple beam test [7]. This design guideline (and others that include FRC) however do not take into account the additional creep that is found for FRC members that are cracked.

Numerous authors have shown that FRC shows increased tensile creep if the member is cracked, e.g. uni-axial tensile creep of Strain Hardening Cement-based Composites (SHCC) [8], uni-axial tensile creep of synthetic macro FRC [9], uni-axial tensile creep of steel FRC [10, 11], flexural creep (e.g. [12, 13]). To the authors' knowledge, except for the publications of Boshoff and Adendorff [8] (micro PVA fibres) and Babafemi and Boshoff [9] (macro synthetic fibres), no investigation has been done on this creep behaviour by applying a creep load to a single fibre embedded in a concrete matrix. Work at this level is needed if the mechanisms behind the increased creep of cracked FRC is to be understood. Once the mechanisms are understood, improvements can be made to the material to reduce this creep and more accurate models to predict the behaviour can be created as it can be based on the fundamental mechanisms.

This paper presents work done on hooked-end steel FRC in order to firstly quantify the tensile creep of this material in a cracked state, and, secondly, to understand the mechanism causing this behaviour. The specific objective of the work is to determine whether matrix damage contributes to the creep behaviour. Tests were done at the macroscale level, i.e. tensile creep tests and also on a single fibre level where rate tests were done by pulling out the fibres at different rates and creep loads were also applied to the embedded fibres individually. Shrinkage tests were also done. Finally, possible mechanisms causing this creep behaviour are identified.

2 Test Setup

2.1 Single Fibre Pull-Out Tests

The single fibre pull-out tests were done in a Zwick Z250 universal testing machine. A 500 kg HBM load cell was used together with two 50 mm HBM LVDTs. The setup can be seen in Fig. 1. The average displacement value of the two LVDTs were used for the pull-out displacement. The specimens were cast by inserting a

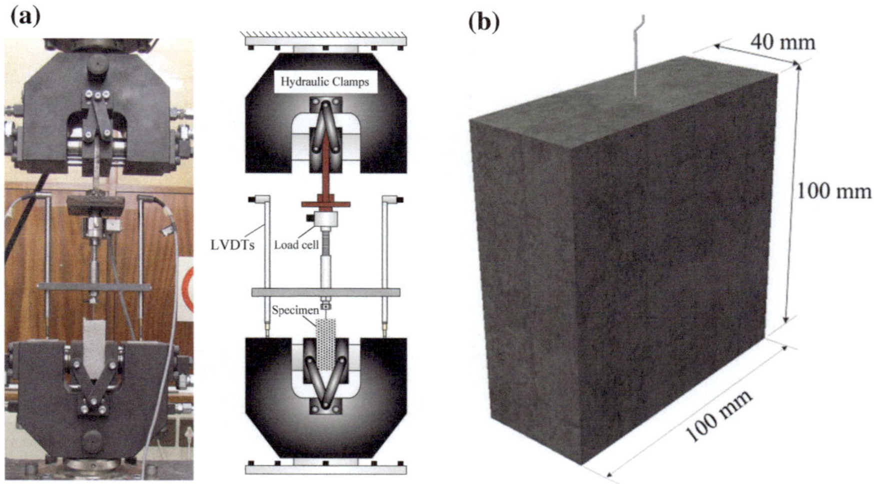

Fig. 1 The single fibre pull-out test setup (**a**), and (**b**), the specimen dimensions

single fibre at a depth of 15 mm in a $100 \times 100 \times 40\ \text{mm}^3$ mould. The sample was vibrated before the fibre insertion and then vibrated lightly again after the fibre was carefully inserted. The mould can be seen in Fig. 1b. The fibres were gripped by drilling a small hole in a bolt and tapped a hole from the side to facilitate a gripping screw. The hooked end of the fibre was cut off before fixing the fibre to the test setup.

2.2 *Single Fibre Pull-Out Creep Test Setup*

The single fibre pull-out creep test setup is different to the setup shown in Sect. 2.1 in order to facilitate a number of samples to be tested simultaneously and not occupying a test machine for the duration of the tests. A fibre was inserted in $100 \times 100 \times 100\ \text{mm}^3$ moulds filled with concrete and the sample was tested upside down in order to facilitate weights to be used to apply the creep load. A threaded rod was fixed to a drilled hole in the bottom of the sample using an appropriate glue and this was in turned fixed upside down to a beam from which all the samples were hanged. The fibre was gripped in a similar fashion as shown in Sect. 2.1 and weights were then applied to the fibre. The setup is shown in Fig. 2. The pull-out displacement was measured by using scaled microscope photos taken periodically. A scaled object was placed next to the fibre for each photo to ensure the calibration is consistently accurate.

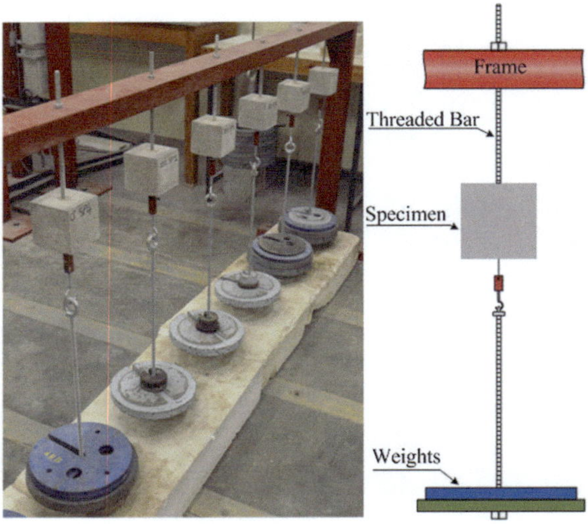

Fig. 2 The single fibre pull-out creep test setup

2.3 Uni-Axial Tensile Test Setup

The uni-axial tensile tests were also done in a Zwick Z250 universal testing machine. Hooks were cast at the two ends of the $100 \times 100 \times 500$ mm^3 specimens to facilitate the gripping of the specimens in the hydraulic grips of the testing machine. A notch with depth of 10 mm was sawn on all four sides of the sample at the centre to ensure the position of the crack. The crack widening was measured using two 10 mm HBM LVDTs connected to two aluminium frames in turn connected to the specimen using a 100 mm gauge length. The test setup is shown in Fig. 3. Note this results in boundary conditions very close to pinned at the top and bottom.

2.4 Uni-Axial Tensile Creep Test Setup

The uni-axial tensile creep tests were done using the same specimens as presented in Sect. 2.3. The load was however applied using two lever arms connected to two samples in series in a loading frame, shown in Fig. 4. A total of eight loading frames were available (i.e. 16 samples could be tested at one point of time) and the ratio of applied weights to applied load in the samples was calibrated using a load cell. This calibration was done prior to the tests with the load cell in the position of the samples and the load cell was then removed before each test. The crack mouth opening was monitored using two 10 mm HBM LVDTs connected to an aluminium frame using a gauge length of 70 mm over the notch, shown in Fig. 4. The readings were taken at regular intervals, ranging from one per second in the

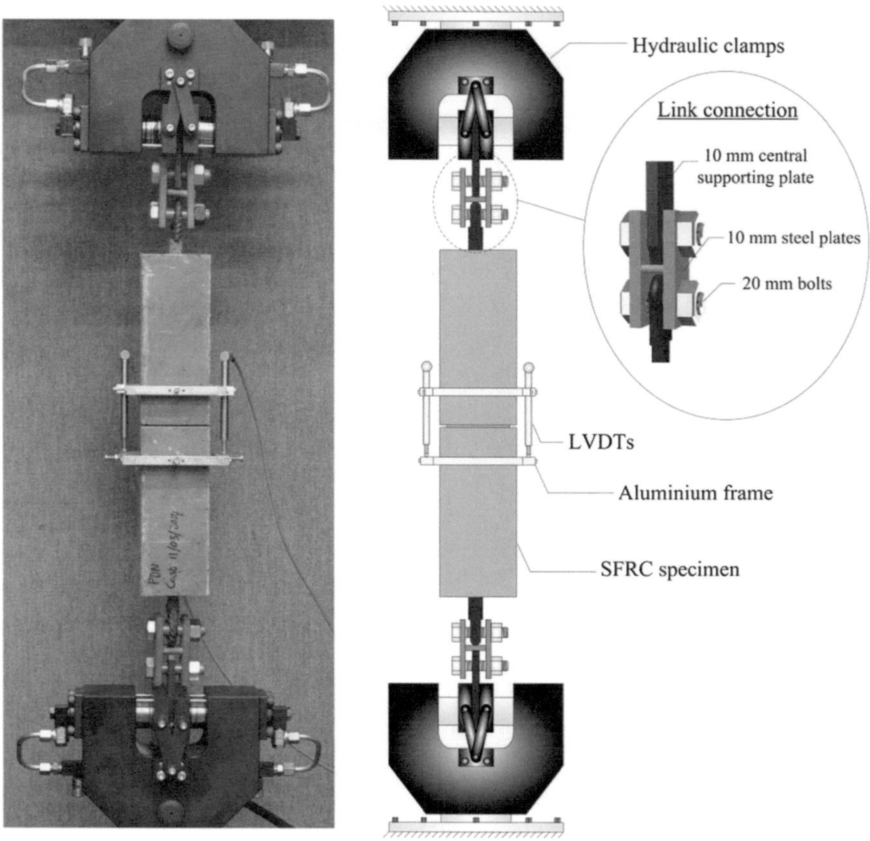

Fig. 3 The uni-axial tensile test setup

beginning to a few a day after a few months. The safety stopper shown in Fig. 4 was also used to apply the load, i.e. the level arms were loaded with the safety stopper in place and the stopper was then removed within a few seconds to apply the load.

2.5 Drying Shrinkage Test Setup

The drying shrinkage was also measured in order to subtract the shrinkage component from the displacement measured over the gauge length in the tensile creep test setup. The setup and results are not discussed in this paper and the reader is referred to [14] for the details. Note that the shrinkage deformations were subtracted from the results shown in this paper.

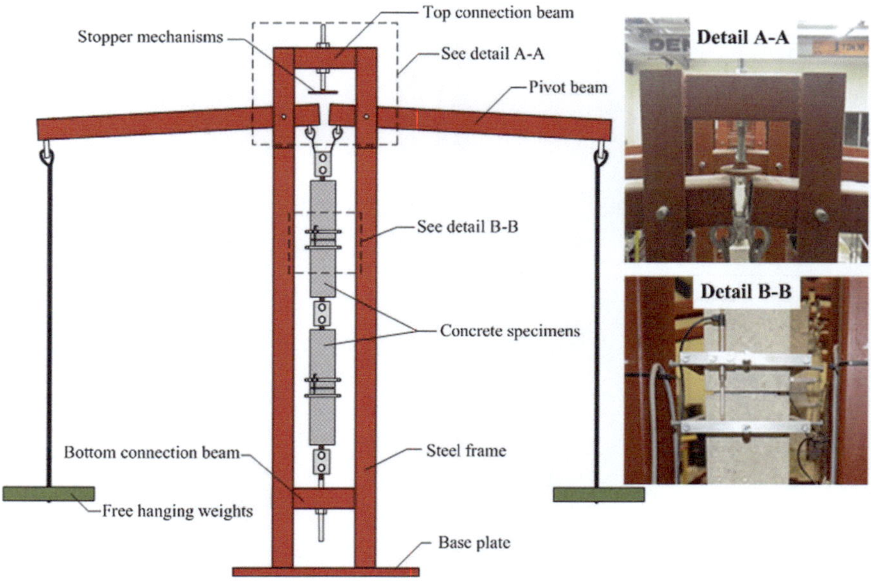

Fig. 4 Uni-axial tensile creep test setup showing the lever arms, the safety stopper and the aluminium frames

3 Materials and Test Programme

3.1 Materials and Mix Design

A concrete with a 60 MPa cube strength was used for the tests. The maximum coarse aggregate size was 6 mm with a water/cement ratio of 0.48. The mix design is shown in Table 1. The steel fibres used were supplied by Bekaert and with the brand name Dramix 3D-65/60-BG. The fibres were 60 mm long with hooked-ends. For the tensile tests, the fibres were added at a volume of 0.5 %. For the single fibre tests the same concrete was used except that the fibres were not added to the matrix. This did result in an increased workability for the concrete used for the single-fibre tests.

Table 1 Concrete mix design

Material type	kg/m^3
Cement (CEM I 52.5)	395
Stone (6 mm nominal size)	800
Sand	990
Water	190
Steel fibre (0.5 % by volume)	39.25
Superplasticiser (0.5 % by weight of binder)	1.975

3.2 Single Fibre Tests

The single fibre pull-out tests were done over 5 orders of magnitude of the pull-out rate in order to determine whether there is a rate-effect in the range of rates tested. The slowest test was done at 0.25 μm/s and rate was increased in multiples of 10 up to 2.5 mm/s. Five tests were done per pull-out rate. The average maximum load over all the rates was used as the reference load (i.e. 100 %) for the single fibre pull-out creep tests.

For the single fibre pull-out creep tests, the fibres were loaded at 30, 50, 70 and 85 % of the average maximum load resisted in the pull-out rate tests. Three specimens were tested for each load level.

Additional tests were also done to verify whether pre-damaging the single fibres would have an effect on the pull-out creep results. In the case for the uni-axial tensile creep tests, where the specimens were pre-cracked, the fibres would have experienced some form of pull-out before the creep loading started. This test was done in order to investigate the effect of this pre-damage on the creep behaviour. Three samples were pre-damaged by pulling out the fibre by 0.53 mm, then unloaded after which the creep load was applied.

Lastly, the samples from the single fibre creep tests were analysed using a micro X-ray computed tomography (μ-CT) scan in order to investigate the damage around the hooked-end of the fibre in the matrix.

3.3 Uni-Axial Tensile Creep Tests

All the specimens used in the uni-axial creep tests were pre-cracked using the setup presented in Sect. 2.3. Each specimen was loaded until a crack width of 0.53 mm and the average stress for the specimens at this crack width was 2.19 MPa. Three specimens were loaded at each load level which are the same as for the single fibre pull-out creep tests, namely 30, 50, 70 and 85 %. Note the actual loading level was calculated according to the specific specimen's residual strength at 0.53 mm crack opening.

4 Test Results

4.1 Single Fibre Pull-Out Rate Tests

These tests were done in order to determine whether there is a rate effect with regards to the pull-out load and also the pull-out displacement at the maximum load. The results for the 0.5 mm/s is shown as an example in Fig. 5a.

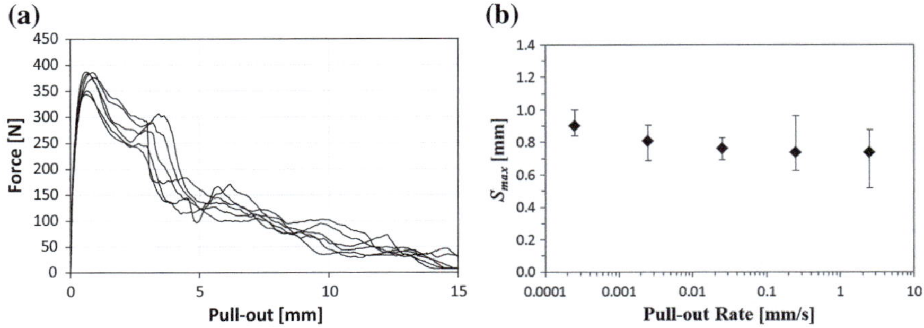

Fig. 5 **a** the pull-out responses of the 0.5 mm/s tests, and **b**, the displacement at the maximum load resisted over the range of pull-out rates tested

The results show there is no change of maximum load resisted of the fibre during pull-out over the pull-out rate range from 0.25 μm/s to 2.5 mm/s. The pull-out displacement at maximum load was however affected by the loading rate and the pull-out at maximum load increased with 22 % from the highest to the lowest pull-out rate as can be seen in Fig. 5b. This gives rise to the hypotheses that stiffness of the interfacial transition zone (ITZ) between the fibre hooked-end and the matrix is affected by the loading rate, but the ITZ does not influence the strength capacity of the hooked-end. One can thus postulate that the ITZ "collapsing" at lower pull-out rates could be one of the causes of the fibre pull-out creep observed in the results shown in this paper.

4.2 Single Fibre Pull-Out Creep Tests

The results of the single fibre pull-out tests are shown in Fig. 6 with (a) the individual results and (b), the average responses. The results are further analysed by

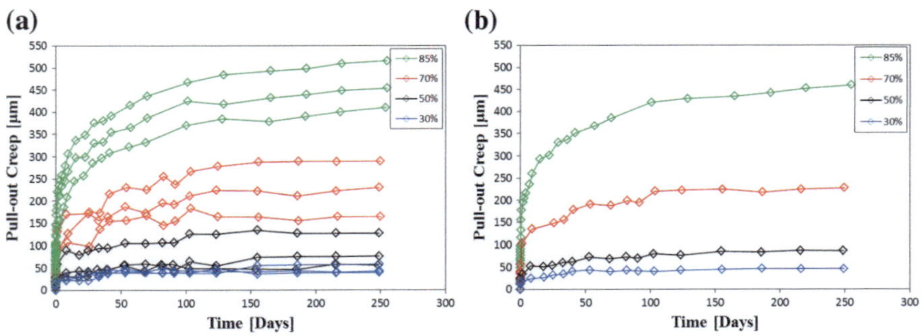

Fig. 6 The individual results of the single fibre pull-out creep tests (**a**), and (**b**), the average responses

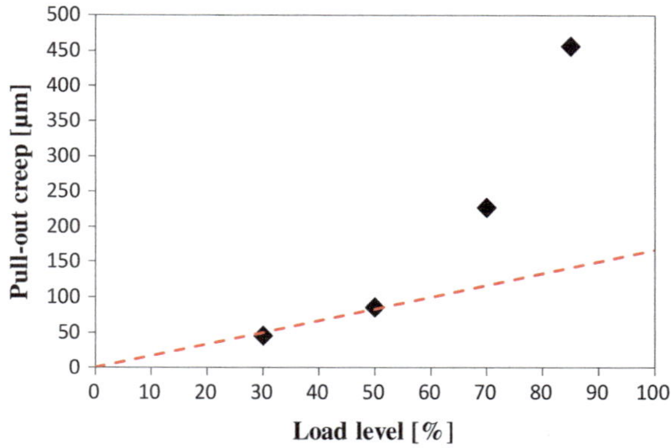

Fig. 7 The total pull-out creep at 240 days plotted against the load level

plotting the total pull-out creep against the applied load level in Fig. 7. It is clear that the pull-out creep is linear to at least 50 % of the load level and then increases non-linearly with an increase of the load level. This indicates that an additional mechanism causing this time-dependant pull-out is possibly activated at loads higher than 50 %.

To investigate this additional mechanism, micro CT scans were taken of the samples at different load levels after 240 days. The focus area was the matrix around the hooked-end of the fibre and the micro CT images are shown in Fig. 8. It is clear from the images that at both 50 and 85 % of the load a gap formed above the hooked-end which is the effect of the pull-out creep. No micro-cracking occurred at the 50 % load, therefore further supported the previously mentioned postulation that the pull-out creep is formed by some form of ITZ crushing and/or

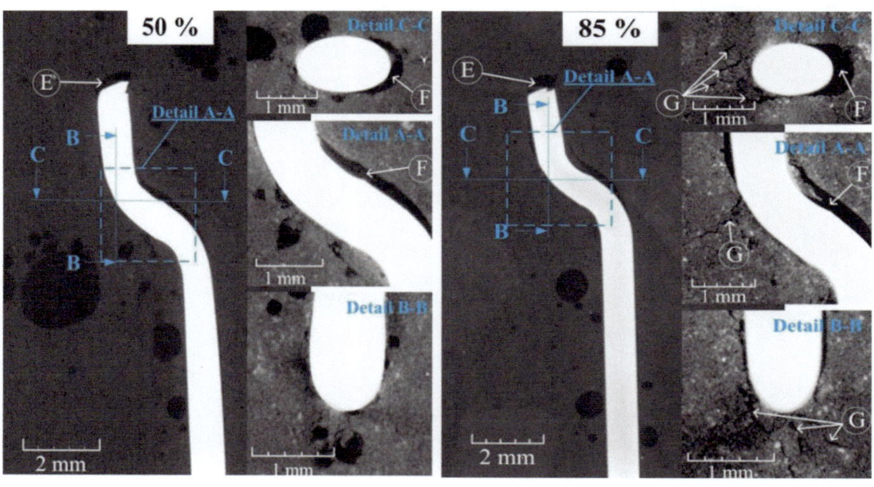

Fig. 8 Micro CT images of a 50 and 85 % specimen after 240 days. The gaps above the hooked-end is indicated by "F" and the micro-cracking is indicated by "G"

collapsing. At the 85 % load level micro-cracking is clearly visible, therefore identifying a second mechanism which causes the fibre pull-out creep. It should be noted that this behaviour is probably not only dependent on the load level, but also on the matrix properties, e.g. w/b ratios, curing etc.

4.3 Uni-Axial Tensile Creep Tests

The individual results of the uni-axial tensile creep tests are shown in Fig. 9 together with the average responses. The total time-dependant crack widening after 240 days is shown in Fig. 10. A similar trend can be seen as for the single fibre pull-out creep tests, i.e. the crack widening being proportional to the load level to at

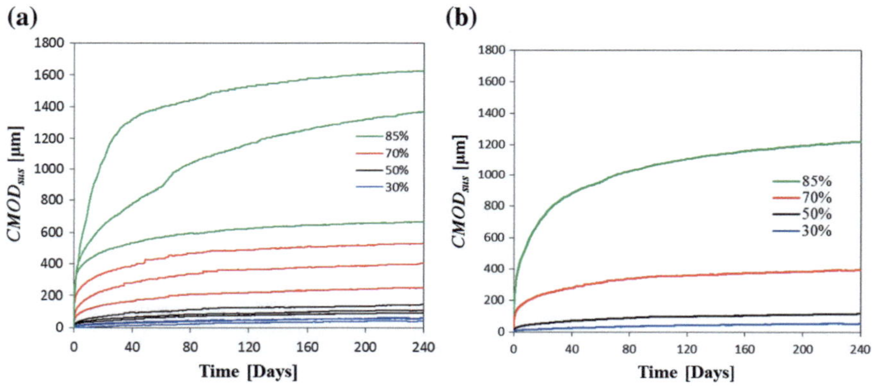

Fig. 9 The individual results of the uni-axial tensile creep tests (**a**), with (**b**), the average responses

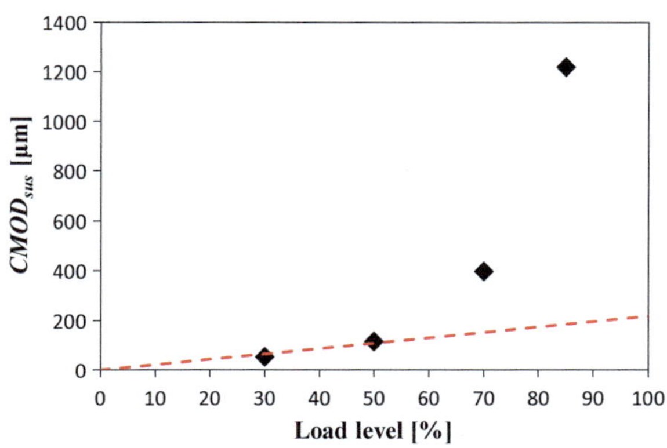

Fig. 10 The total time-dependant crack widening after 240 days for the uni-axial tensile creep tests

least to the 50 % load level. It is also interesting to note that the time-dependant crack widening is around double the single fibre pull-out creep, which is to be expected as in the uni-axial tensile creep, the fibres creep from both sides. This confirms the mechanism of creep in the single fibre pull-out creep tests is similar to the uni-axial tensile creep tests.

4.4 Pre-Damaged Single Fibre Pull-Out Creep Tests

In order to look at the effect of the crack formation, which causes some initial fibre pull-out, on the single fibre pull-out tests, samples were pull-out out to 0.53 mm before commencing of the single fibre pull-out creep tests at a 50 % load level. The average responses of both the virgin 50 % tests and the pre-damaged creep tests are shown in Fig. 11, in both linear and logarithmic scales.

It is clear that the total pull-out creep is the same, whether the fibre was pre-damaged or not. However, it can be seen on Fig. 11b that the responses deviates substantially up to about 30 days. This can be explained using the postulation that a ITZ causes a significant part of the creep. For the pre-damaged fibre, the ITZ was crushed during the pre-test damage/pull-out and thus prevented the initial creep as was observed by the undamaged specimens. After time, this became insignificant and the results of the two sets of samples converged.

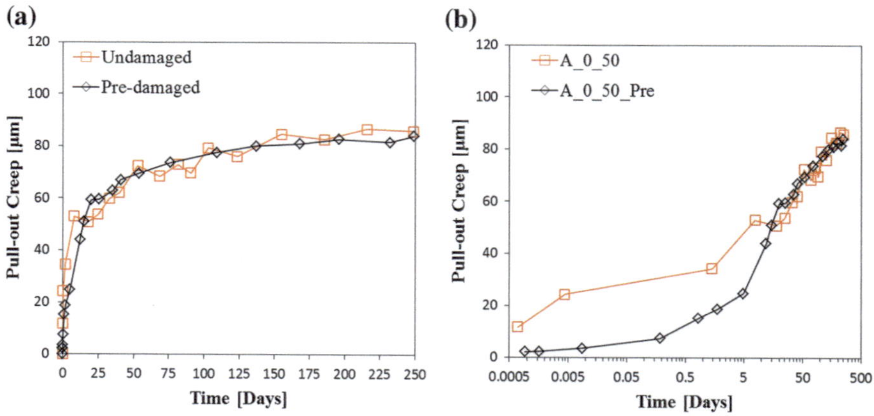

Fig. 11 The average responses of the single fibre pull-out results of the 50 % load level for both the virgin and the pre-damaged samples (**a**), and (**b**), the same results on a log scale

5 Conclusions

In this paper, results are reported of a study which investigated the mechanism causing the time-dependant crack widening of cracked steel fibre reinforced concrete members. Rate and creep tests were done by pulling out single fibres and uni-axial tensile creep tests were also done. The mechanisms causing this time-dependant deformation is not yet fully understood, however, the following conclusions can be made based on the work presented here:

- Single fibre pull-out tests can be used to investigate the creep behaviour of cracked steel fibre reinforced concrete.
- For the range tested, it was found that the pull-out rate does not affect the maximum pull-out resistance of hooked-end steel fibres. However, the pull-out displacement at the point of maximum pull-out resistance was increased by 22 % over five orders of magnitude of the pull-out rate.
- Both the single fibre pull-out creep results and the uni-axial tensile creep results showed that the time-dependant displacements are proportional to a load level of at least 50 % of the ultimate resistance after which it increases non-linearly.
- It was shown that micro-cracking is responsible for this non-linear increase of pull-out creep.
- Based on the rate test results of the single fibre pull-out, the micro-CT images and the pre-damaged single fibre pull-out creep tests it can be deduced that the interfacial transition zone between the fibre hooked-end and the matrix plays and important role in the creep of cracked fibre reinforced concrete.

References

1. Thomas, F.G.: Creep of concrete under load, pp. 292–294. International Association of Testing Materials, London Congress (1937)
2. Pickett, G.: The effect of change in moisture content on the creep of concrete under a sustained load. J. ACI **38**, 333–355 (1942)
3. Wendner, R., Hubler, M.H., Bažant, Z.P.: Statistical justification of model B4 for multi-decade concrete creep using laboratory and bridge databases and comparisons to other models. Mater. Struct. **48**, 815–833 (2015)
4. EN 1992-1-1: Eurocode 2: Design of Concrete Structures—Part 1-1: General Rules and Rules for Buildings. Brussels, European Committee for Standardization (2004)
5. Brandt, A.M.: Fibre reinforced cement-based (FRC) composites after over 40 years of development in building and civil engineering. Compos. Struct. **86**, 3–9 (2008)
6. *fib* Special Activity Group 6, 2012, Model Code 2010, *fib* bulletin 65 and 66
7. EN 14651: Test method for metallic fibre concrete—measuring the flexural tensile strength (limit of proportionality (LOP), residual), Eurocode. Brussels, European Committee for Standardization (2005)
8. Boshoff, W.P., Adendorff, C.J.: Effect of sustained loading on SHCC crack widths. Cement Concr. Compos. **37**, 119–125 (2013)

9. Babafemi, A.J., Boshoff, W.P.: Tensile creep of macro-synthetic fibre reinforced concrete (MSFRC) under uni-axial tensile loading. Cement Concr. Compos. **55**, 62–69 (2015)
10. Mouton, C.J., Boshoff, W.P.: Initial study on the tensile creep of cracked fibre reinforced concrete. In: Proceedings of BEFIB. Guimaraes, Portugal (2012)
11. Zhao, G., di Prisco, M., Vandewalle, L.: Experimental investigation on uniaxial tensile creep behavior of cracked steel fiber reinforced concrete. Mater. Struct. **48**, 3173–3185 (2015)
12. Tan, K., Saha, M.: Ten year study on steel fibre-reinforced concrete beams under sustained loads. ACI Struct. J. **102**, 472–480 (2005)
13. Zerbino, R., Barragan, B.: Long-term behaviour of cracked steel fibre-reinforced concrete beams under sustained loading. ACI Mater. J. **109**, 215–224 (2012)
14. Nieuwoudt, P.D.: Time-dependent behaviour of cracked steel fibre reinforced concrete: from single fibre level to macroscopic level. Ph.D. Dissertation, Stellenbosch University, South Africa (2016)

Flexural Post-cracking Creep Behaviour of Macro-synthetic and Steel Fiber Reinforced Concrete

P. Pujadas, Ana Blanco, Sergio H.P. Cavalaro, Albert de la Fuente and A. Aguado

Abstract In this paper the post-cracking creep behaviour of FRC beams under flexural load is evaluated in order to determine whether under certain loading conditions plastic fibres may be safely used in the long-term without compromising the serviceability requirements. For that, an experimental program was conducted that involved the testing of 30 beams with dimensions of 150 × 150 × 600 mm reinforced either with plastic or steel fibres. The creep test setup consisted in a four-point bending test in previously cracked beams up to crack widths of 0.25, 1.50 and 2.50 mm. The sustained load ranged between 50 and 60 % of the cracking load and was applied by means of a lever system. The sustained load was controlled throughout the test with a load cell. The tests were performed under two different environmental conditions during 6 months. Despite the large deformations exhibited by plastic fibres under sustained load over time, their use as reinforcement should not be discarded as long as the effects of creep are considered in the design.

Keywords Plastic fibre · Steel fibre · Time dependent behaviour · Creep · Secondary creep · Tertiary creep

1 Introduction

Steel fibres are still preferably used over macro-synthetic fibres. One of the most recurrent reasons for that is its creep behavior which is the time-dependent strain that develops in concrete due to sustained stress. According to [1], in fibre reinforced concrete (FRC) it is not possible to uncouple the time dependent behavior of the concrete from that of the fibre reinforcement. In this sense, [2, 3] define the mechanism of creep in bending of a cracked element as the result of: (1) concrete creep in compression, (2) the time dependent debonding behavior at fibres-matrix interface and (3) the fibres creep at material level in tension. Thus, the latter is

P. Pujadas (✉) · A. Blanco · S.H.P. Cavalaro · A. de la Fuente · A. Aguado
Universitat Politècnica de Catalunya (UPC-Barcelona Tech), Barcelona, Spain
e-mail: pablo.pujadas@upc.edu

P. Serna et al. (eds.), *Creep Behaviour in Cracked Sections of Fibre Reinforced Concrete*, RILEM Bookseries 14,
DOI 10.1007/978-94-024-1001-3_7

significantly susceptible to the fibre material and whereas in SFRC, the postcracking creep phenomena is not caused by the creep deformation of the actual fibres, but rather by the slow pullout of the fibres from the matrix [4], in PFRC, the creep behaviour of synthetic fibres is a complex phenomenon which depends not only on the stress and temperature [5] but also on intrinsic material properties such as the crystallinity and the molecular orientation as well as other external parameters such UV and moisture [6]. However, the extent to which the aforementioned factors affect the performances of PFRCs is still not completely clear and further experimental investigations are needed.

Despite the relevance of the issue, flexural creep behavior of FRC is still a controversial issue that requires thorough research, particularly in the case of plastic fibre reinforced concrete (PFRC) since it may exhibit additional creep due to the fibres. The main goal pursued in this paper is to evaluate the post-cracking creep behaviour of FRC beams under flexural load and to determine whether under certain loading and crack-width conditions plastic fibres may be safely used in the long-term without compromising the serviceability requirements of the structures.

2 Experimental Method

To investigate the FRC time-dependent behavior, a number of 30 flexural beam tests (with dimensions of 150 × 150 × 600 mm) reinforced either with plastic fibres (PF) or steel fibres (SF) were carried out following a 4-point bending test. The test were done in two different environmental conditions (S1 and S2) during 5 months considering different precrack-widths and load levels.

2.1 Materials and Mix Design

Four series of concrete were produced, one with 5 kg/m^3 of PF and another 40 kg/m^3 of SF for each of the two environmental conditions (S1 and S2). The details of the concrete mix for all series are presented in Table 1.

The plastic macro-fibres (PF) used in the tests were derived from polyolefin with a continuously embossed surface texture to improve adherence and they had straight shape and a rectangular cross-section. The steel fibres (SF) were made of low carbon steel and are gathered into bundles by water-soluble glue and had a circular cross section and hooked ends. Further properties are presented in Table 2.

Table 3 shows the average compressive strength (f_{cm}), average modulus of elasticity (E_{cm}), limit of proportionality (f_L) and the residual flexural tensile strengths (f_{R1}, f_{R2}, f_{R3} and f_{R4}) corresponding to CMODs of 0.05, 0.50, 1.50, 2.50 and 3.50 mm, respectively.

Table 1 Composition of the FRC mixture (in kg/m^3)

Materials	Characteristics	Plastic fibres		Steel fibres	
		S1_PF	S2_PF	S1_SF	S2_SF
Gravel (6/15 mm) (kg/m^3)	Granite	520	520	520	520
Gravel (2.5/6 mm) (kg/m^3)	Granite	400	400	400	400
Sand (0/3 mm) (kg/m^3)	Granite	500	500	500	510
Cement (kg/m^3)	CEM I 52,5 R	400	400	400	350
Filler (kg/m^3)	Marble dust	260	260	260	300
Water (kg/m^3)	–	170	178	168	178
Superplasticizer (kg/m^3)	Adva® flow 400	12	12	12	12
Fibres (kg/m^3)	PF/SF	5 (PF)	5 (PF)	40 (SF)	40 (SF)

Table 2 Fibre characteristics (data provided by the manufacturer)

		PF	SF
Length (*L*)	(mm)	48	50
Equivalent diameter (*d*)	(mm)	–	0.62
Aspect ratio (*L/d*)	(–)	–	83
Tensile strength (f_y)	(MPa)	550	1270
Modulus of elasticity (*E*)	(GPa)	10	210
Number of fibres per kg	(fibras)	>35,000	8100

2.2 *Test Procedure*

The specimens tested were subjected to two different stages (see scheme in Fig. 1): phase 1 or pre-cracking and phase 2 or long term loading test which was extended up to the unloading and recovery. In the first phase the specimens were tested in a closed-loop servo-hydraulic system according to a four-point bending with a 450 mm span and 25 mm notch, using crack mouth opening displacement (CMOD) as the control signal. In addition to the clip gauge placed at the bottom of the beam, the crack-opening displacement (w_p^f) was monitored through a linear variation differential transducer (LVDT) fixed on a lateral side of the specimen 12 mm above the bottom surface of the beam. The process was interrupted at different w_p^f levels ($w_p^f = 0.25 - 1.50 - 2.50$). Table 4 shows the w_p^f of each of the beams tested.

The pre-cracked beams were then unloaded and immediately after placed in a creep frame. The design of the test set up for the long term flexural test were based on the previous work by Bast et al. [7] and basically identical to that used by [4, 8, 9]. The steel frame used allows to simultaneously perform the long term test in various specimens set in a column. In this case, three specimens were loaded in each column, under a 4-point bending test configuration using steel rollers as a free supports (see Fig. 2). The lateral LVDT used to measure the w_p^f during the first stage (which always remained settled to the beam) was used for the control of the crack evolution.

Table 3 Characterization of the FRC at 28 days

	S1_PF		S2_PF		S1_SF		S2_SF	
	Average (MPa)	CV (%)	Average (MPa)	CV (%)	Average (MPa)	CV (%)	Average (MPa)	CV (%)
E_{cm}	31,150	1.69	–	–	31,597	1.08	30,160	2.20
f_{cm}	52.15	1.52	48.89	1.57	54.30	1.51	46.77	2.54
f_L	4.61	2.19	4.22	2.66	3.73	8.57	3.76	7.96
$f_{R,1}$	2.01	22.30	2.38	15.54	4.62	12.15	3.75	22.29
$f_{R,2}$	2.25	28.53	2.93	20.89	5.09	13.77	4.24	17.91
$f_{R,3}$	2.46	26.84	3.32	24.15	5.10	15.91	4.30	15.88
$f_{R,4}$	2.48	23.47	3.49	27.40	4.87	14.08	4.17	15.68

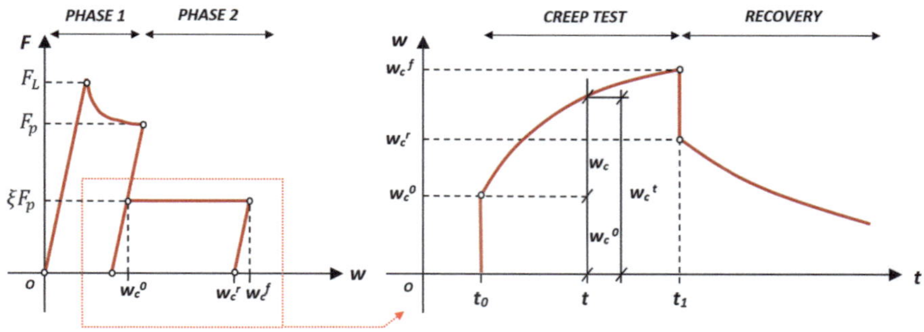

Fig. 1 Diagram of the complete creep test procedure

Table 4 Variation in environmental conditions over the test days

S1	Apr	May	Jun	Jul	Aug	Sep
Humidity (%)						
Max.	59.6	56.4	73.7	78.7	80.7	81.3
Min.	48.2	43.2	36.3	60.3	73.1	53.1
Average	52.7	50.4	57.9	69.1	78.3	73.3
Temperature (°C)						
Max.	22.3	24.9	25.7	26.4	26.6	26.7
Min.	20.8	20.8	23.3	23.7	24.5	24.3
Average	21.3	21.8	24.5	24.7	25.5	25.8

S2	Feb	Mar	Apr	May
Humidity (%)				
Max.	78.0	78.8	65.5	73.0
Min.	12.3	16.5	27.3	34.8
Average	36.9	49.9	48.8	50.3
Temperature (°C)				
Max.	19.2	19.7	24.1	25.7
Min.	6.5	9.1	15.3	17.1
Average	12.7	15.3	19.2	21.3

The humidity and temperature in the S1 and S2 conditions of the experimental program were different. Table 4 shows the variation in environmental conditions over the test duration.

In the S1 the specimens were tested in a climate-controlled room under relatively constant conditions of humidity and temperature (Table 4). The S2 beams were exposed to ambient conditions with alternating humidity and temperature (as observed in Table 4). The theoretical load levels considered in the study were between 50 and 60 % of the load F_p, registered when the pre-cracking stage was stopped (see Table 5).

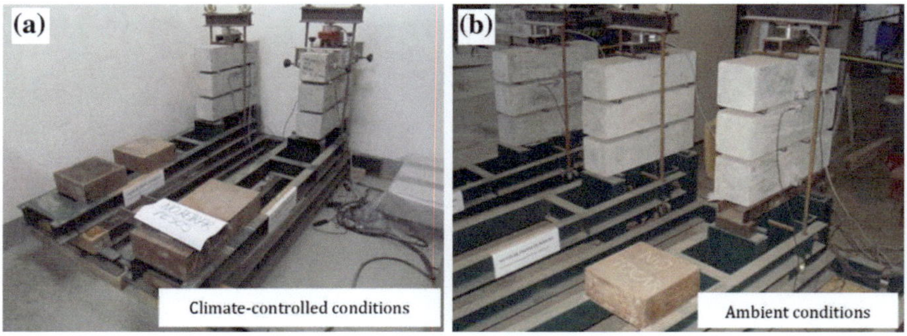

Fig. 2 Creep frames during the specimens loading **a** S1 and **b** S2

Table 5 Summary of loading conditions and pre-crack-width

		Specimens	w_p^f (mm)	F_c/F_p (%)
S1	PF	S1/PF_0.25P1	0.25	48.9
		S1/PF_0.25P2		58.2
		S1/PF_0.25P3		69.1
		S1/PF 0.25P4		57.3
		S1/PF_0.25P5		60.5
		S1/PF_1.50P6	1.50	50.0
		S1/PF_1.50P7		50.4
		S1/PF_2.50P8	2.50	54.0
		S1/PF_2.50P9		48.6
	SF	S1/SF_0.25P1	0.25	53.9
		S1/SF_0.25P2		61.6
		S1/SF_0.25P3		69.8
		S1.SF_1.50P4	1.50	52.9
		S1/SF_2.50P5	2.50	49.5
		S1/SF_2.50P6		47.5
S2	PF	S2/PF_0.25P1	0.25	48.9
		S2/PF _0.25P2		48.3
		S2/PF_0.25P3		48.6
		S2/PF_0.25P4		52.6
		S2/PF_1.50P5	1.50	43.1
		S2/PF_2.50P6	2.50	49.3
	SF	S2/SF_0.25P1	0.25	47.0
		S2/SF_0.25P2		62.2
		S2/SF_0.25P3		52.3
		S2/SF_0.25P4		52.3
		S2/SF_0.25P5		46.5
		S2/SF_0.25P6		53.7
		S2/SF_1.50P7	1.50	46.8
		S2/SF_1.50P8		48.0
		S2/SF_2.50P9	2.50	37.1

3 Experimental Results and Discussion

3.1 Creep Crack Width Results

For the case of the S1 beam, after 150 days of loading the values of Δw_c^t (increment of crack width) registered for the specimens precracked at w_p^f of 0.25 mm vary between 0.30 and 0.45 mm for PF (see Fig. 3a) and between 0.08 and 0.13 for SF (see Fig. 3b). The specimens with a w_p^f of 1.50 mm, reached values of Δw_c^t between 1.20 and 1.40 mm for PF (see Fig. 3c), whereas SF reached only 0.15 mm (see Fig. 3d).

The creep curve of figure comprises two stages: primary and secondary creep. The increase of Δw_c^t is relatively high the first days just right after the sustained load is applied (primary creep), subsequently decreases progressively (secondary creep). On average, about half (50 %) of Δw_c^t recorded after 150 days of the test occurs in the first 5 days, after 30 days 70 % and after 90 days 90 % of the final Δw_c^t is reached.

In the case of the S2 beams (under ambient environmental conditions) after 90 days of testing the values Δw_c^t obtained vary between 0.30 and 0.45 mm for the specimens precracked with w_p^f of 0.25 mm (see Fig. 4a). The S2/PF_1.50P5 pre-cracked with a w_p^f of 1.50 mm reach approximately 0.90 mm (see Fig. 4b).

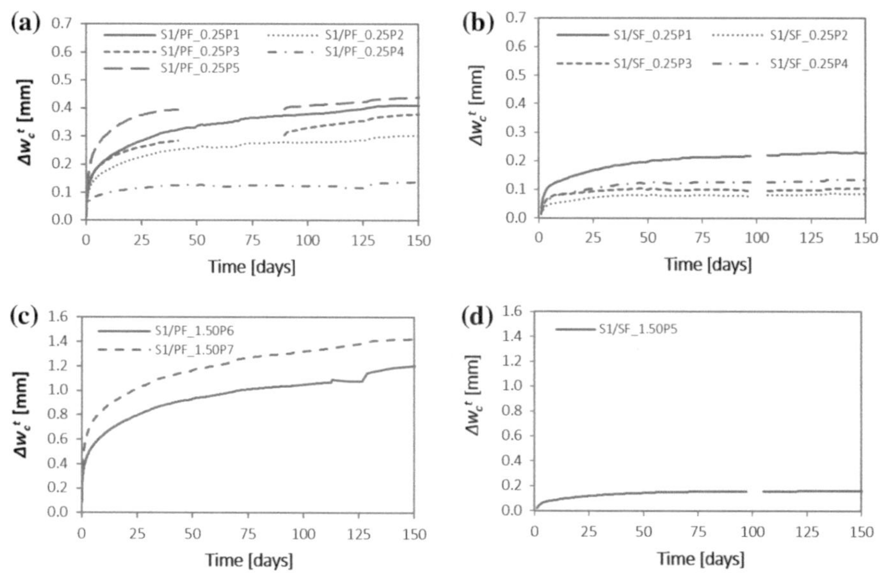

Fig. 3 Diagram Δw_c^t – *time* curves for the specimens: **a** S1/PF_0.25Pi; **b** S1/SF_0.25Pi; **c** S1/PF_1.50Pi; **d** S1/SF_1.50Pi

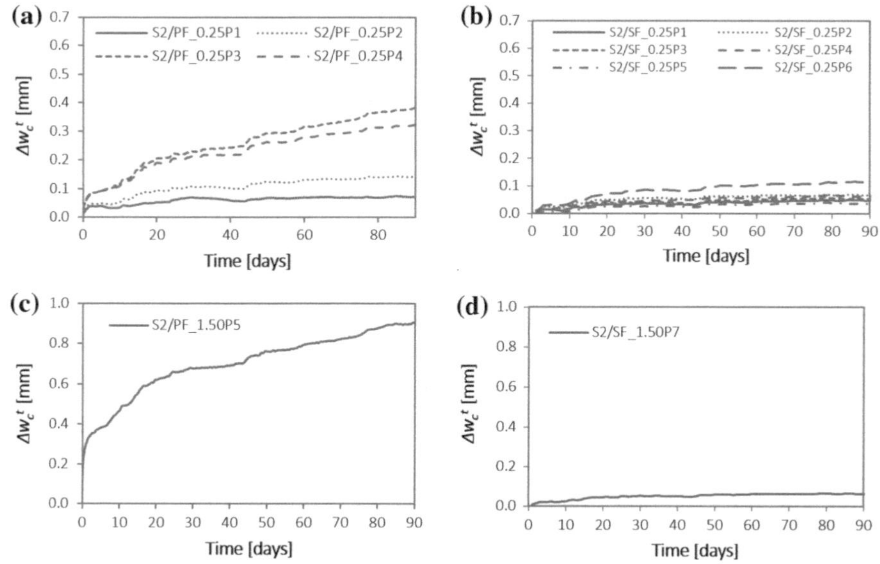

Fig. 4 Diagram $\Delta w_c^t - time$ curves for the specimens: **a** S2/PF_0.25Pi; **b** S2/SF_0.25Pi; **c** S2/PF_1.50Pi; **d** S2/SF_1.50Pi

Notice that the nature of the S2 creep curve is considerably different from that of S1. In the S2 specimens, creep increases monotonously in time, while in the S1 specimens creep stabilizes within several days. However, the initial rate of the S1-creep curve is larger. Unexpectedly during some initial period (about 7–8 days in our tests) the S2 creep width is less than the S1 creep for equivalent specimens. Only then the S2 creep begins to exceed the S1 creep.

During the execution of the test, none of the beams previously analysed (with w_p^f of 0.25 and 1.5 mm) entered into the tertiary creep stage (a period of increasing creep rate, prior to failure). However, for all those beams with a w_p^f of 2.50 mm a three-stage creep response: primary, secondary, and tertiary creep with a sudden failure was observed (see Fig. 5). In this experimental campaign, failure took place after 41, 112 and 108 days of loading for S1/PF_2.50P8 (with load ratio of $\%F_p = 54\ \%$), S1/PF_2.50P9 ($\%F_p = 48\ \%$) and S2/PF_2.50P6 ($\%F_p = 49\ \%$) respectively. On the contrary, none of the SF beams with a w_p^f of 2.50 mm entered in tertiary creep. The values of Δw_c^t recorded range between 0.1 and 0.2 mm, lower values than those obtained for PF with a w_p^f of 0.25 mm.

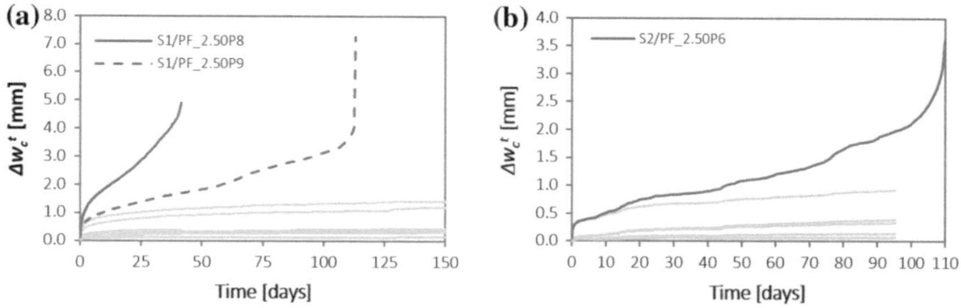

Fig. 5 Diagram Δw_c^t – *time* curves for the specimens: **a** S1/PF_2.50Pi; **b** S2/PF_2.50Pi

3.2 Creep Coefficient Results

Considering the inherent scatter of creep, and the intrinsic scatter of the material together with the fact that each specimen was tested under different conditions of width and load, it is difficult to draw obvious conclusions from a direct analysis of the results. To overcome these drawbacks, the analysis of the creep coefficient is proposed.

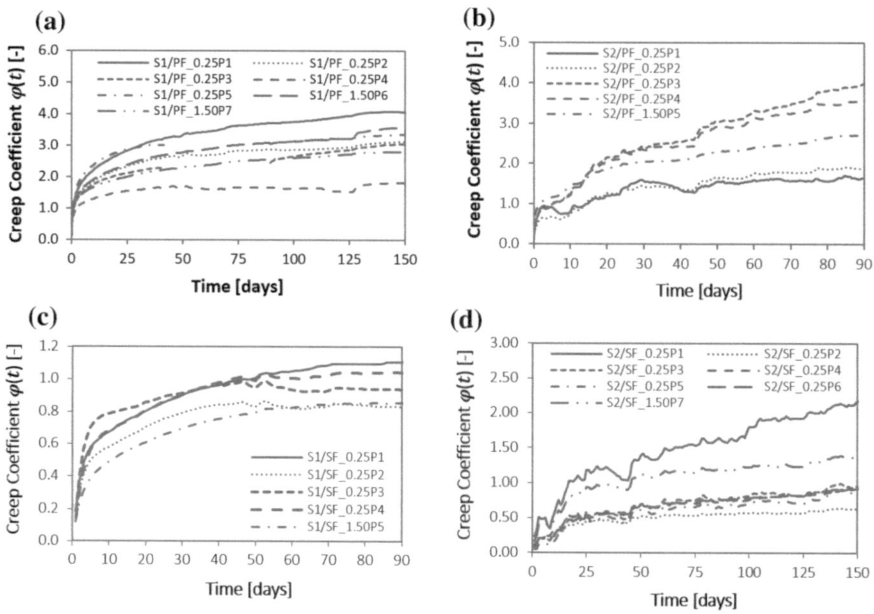

Fig. 6 $\varphi(t)$ – *time* curves of the specimens: **a** S1/PF_0.25Pi and 1.50Pi; **b** S2/PF__0.25Pi and 1.50Pi; **c** S1/SF_0.25Pi and 1.50Pi; **d** S2/SF_0.25Pi and 1.50Pi

For the purpose of this research, the creep coefficient ($\varphi(t)$) at a time t defined in terms of crack width is the ratio between the crack at t time t in phase 2 (w_c^t) and the initial crack width (w_c^o), as indicated in expression:

$$\varphi(t) = \frac{w_c^t}{w_c^o} \quad (1)$$

The evolution of the creep coefficient is presented in Fig. 6 for each beam. The curves are grouped by stages S1/PF (Fig. 5a), S2/PF (Fig. 5b), S1/SF (Fig. 5c) or S2/SF (Fig. 5d).

Using t = 90 days as reference, a value of $\varphi(t = 90)$ between 0.5 and 4.0 is reached for all the specimens of both conditions S1 and S2. PF creep coefficients for 90 days ($\varphi^{t \rightarrow 90}$) ranged between 1.5 and 4.0 for both S1 and S2 campaigns, although in the former an asymptotic value of the creep coefficient was reached during the secondary stage. In this stage, creep is not significantly affected by the w_p^f value. Moreover, at similar loading level and for $w_p^f \leq 1.50$ cracked PFRC can be expected to experience creep coefficients of around twice those of SFRC.

4 Conclusions

The largest crack widths during secondary creep were obtained for those specimens with higher pre-cracked width. Such outcome is reasonable since the pull-out and the tensile creep of the plastic fibre is accentuated with the load level and the cracking level. It is important to highlight that although creep rate of plastic fibre is twice the value for steel fibres, all along secondary creep, this rate is not affected by the pre-cracked width value. Such fact may be advantageous in order to predict the creep behaviour in secondary creep, and meet the design crack widths criterion.

However, for advanced cracking levels the reinforcing action of plastic fibres is no longer effective due to the significant tensile creep deformation they suffer (although the fibres did not break in any of the analysed cases). Consequently, the concrete crack progresses and reduces the compression zone until it is depleted, causing the tertiary creep and consequently the failure of the specimens. This failure mode was not observed in the SFRC beams, even for those specimens with a pre-cracked width of 2.5 mm.

Despite the large deformations exhibited by plastic fibres under sustained load over time, their use as reinforcement should not be discarded as long as the effects of creep are considered in the design as it is indicated in the Model Code 2010. For that, design criteria defining the scope of application of fibre reinforcement, particularly in the case of plastic fibres, needs to be established to ensure the ULS and SLS requirements. The secondary creep should be considered as a design criterion of prime importance in order to determine the life of a given PFRC structure. For that, it should be ensured that with adequate probability, cracks will not impair the functionality, durability and appearance of the structure and define or limit the range of applications of this material.

References

1. Bernard, E.S.: Influence of fiber type on creep deformation of cracked fiber-reinforced shotcrete panels. ACI Mater. J. **107**(107), 474–480 (2010)
2. Boshoff, W.P., Mechtcherine, V., van Zijl, G.P.A.G.: Characterising the time-dependant behaviour on the single fibre level of SHCC: Part 1: Mechanism of fibre pull-out creep. Cem. Concr. Res. **39**(9), 779–786 (2009)
3. Abrishambaf, A., Barros, J.A.O., Cunha, V.M.C.F.: Time-dependent flexural behaviour of cracked steel fibre reinforced self-compacting concrete panels. Cem. Concr. Res. **72**, 21–36 (2015)
4. Zerbino, R.L., Barragán, B.E.: Long-term behavior of cracked steel fiber-reinforced concrete beams under sustained loading. ACI Mater. J. **109**(109), 215–224 (2012)
5. Buratti, N., Mazzotti, C.: Experimental tests on the effect of temperature on the long-term behaviour of macrosynthetic fibre reinforced concretes. Constr. Build. Mater. **95**, 133–142 (2015)
6. Liu, G., Huang, X., Deng, Y., Wang, C., Tong, X., Liu, W., Huang, Y., Yang, J., Liao, Q., Li, X.: Study on the creep behavior of polypropylene Xiaolin. Polym. Eng. Sci. **49**(7), 1375–1382 (2009)
7. Bast, T., Eder, A., Regensburg, F., Bauingenieurwesen, F.: Untersuchungen zum Langzeitverhalten von Faserbetonen unter Biegezugbeanspruchung—ein Zwischenbericht. No. Bild **2**, 32–35 (2005)
8. Arango, S.E., Serna, P., Martí-Vargas, J.R., García-Taengua, E.: A test method to characterize flexural creep behaviour of pre-cracked FRC specimens. Exp. Mech. **52**(8), 1067–1078 (2011)
9. García-Taengua, E., Arango, S., Martí-Vargas, J.R., Serna, P.: Flexural creep of steel fiber reinforced concrete in the cracked state. Constr. Build. Mater. **65**, 321–329 (2014)

Part III
Creep on Special Materials

Behavior of Cracked Cross-Section of Fibre Reinforced UHPFRC Under Sustained Load

Daniele Casucci, Catherina Thiele and Jürgen Schnell

Abstract In this paper, the work done at TU Kaiserslautern for a better understanding of the behaviour of cracked ultra-high performance fibre reinforced concrete (UHPFRC) is reported. UHPC is a material with a very high compressive strength. A ductile behaviour and a rather good tensile strength are achieved by addition of fibres. While the short time resistance can be easily evaluate with laboratory tests, the discussion concerning the long-term behaviour is still open. A sustained load test rig with lever arms for axial tension tests was built. Dog-bone specimen and prisms were loaded for axial and bending creep tests investigating many different parameters. Aim of this research is to give a systematic characterisation of the problem and find out if the material can creep up to failure and at which load level. Required consequences for design of UHFRC should be defined. The computed tomography was used for a direct observation of the fibre inside the concrete. A correlation between experimental results with large scatter and specimen fibre distribution and orientation is sought in this way.

Keywords UHPFRC · Steel microfibres · Tensile creep · Computed tomography

1 Introduction

UHPFRC is a kind of evolution of the ordinary concrete, every ingredient of the mixture was optimized to reach a micro-structure with maximum density possible. The most appreciated material property is an outstanding compressive strength, (150–250 MPa). Disadvantages of UHPC are the relative poor tensile strength and the brittle fracture behaviour under compressive and tensile loads. A ductile material behaviour, higher tensile strength and improved crack width control have been achieved by adding fibres. UHPFRC is indeed usually reinforced with steel

D. Casucci (✉) · C. Thiele · J. Schnell
Institute of Concrete Structures and Structural Engineering, TU Kaiserslautern, Kaiserslautern, Germany
e-mail: daniele.casucci@bauing.uni-kl.de

P. Serna et al. (eds.), *Creep Behaviour in Cracked Sections of Fibre Reinforced Concrete*, RILEM Bookseries 14,
DOI 10.1007/978-94-024-1001-3_8

micro fibres, which have in most of the cases a length of 5–20 mm and a diameter of 0.1–0.2 mm [1]. The fibre content is generally higher compared to ordinary fibre reinforced concrete. That corresponds to about 160 kg/m^3 (2 %) and usually originates, as described below, the so-called “strain hardening behaviour”.

Since the UHPFRC exhibits excellent properties, it seems reasonable to reduce or in some cases even avoid conventional reinforcement bars. The aim of this research work is to investigate the long term material behaviour of cracked UHFPRC. Which kind of deformations have to be expected? What happens to the fibre adhesion over the time? What kind of progressive pullout may threat the load—bearing capacity of structural UHFPRC elements?

First of all, it is important to remark that the geometry of the fibres is different: generally UHPC fibres are not hocked or undulated. The anchorage is given only by the chemical bond between cement and fibre combined with micro-roughness of the fibres.

While the literature concerning this phenomenon on ordinary reinforced concrete, is quite poor, that one concerning UHPC is very rare. Most of the research is indeed oriented to the tensile creep in undamaged cross sections and in early age stage.

Candido, Vasanelli et al. [2, 3] investigated the behaviour of ordinary fibre reinforced concrete combining reinforcement bars with hooked and waved steel fibres. Full scale beams under environmental exposure were subjected to a sustained load over 72 months. A stabilization of the crack width after aprox. 10 months was found. They confirmed the influence of the fibres for crack width limitation and the positive effect for the stiffness of the beams.

Zhao et al. [4] investigated pre-cracked hocked fibre reinforced concrete specimen with (fibre content 80 kg/m^3) with crack widths of 0.05 and 0.20 mm. At first, the specimens were loaded with 30 % of the ultimate crack load for a three month period and with 60 % of the crack load afterwards. In case of the 0.05 mm crack width, the width remained almost constant during the whole observation. An insignificant recovery of the plastic deformations was observed. Moreover, at 30 % pre-cracking load level, no loss of stiffness and no damage progression was detected. However, investigations are missing for higher load levels.

Arango et al. [5] observed that the slenderness ratio of the fibres in conjunction with the fibre aggregate size have an influence on the bending creep behaviour of short-fibre reinforced concrete (SFRC). He performed three point bending tests on pre-cracked specimens with a 0.5 mm crack width. After the crack formation the specimens were totally unloaded and re-loaded with a sustained load for 90 days. The specimen were applied with 60 and 80 % of the maximum crack load. Bernard [6] assessed that steel fibres are less sensitive to tensile creep than macrosynthetic fibres and observed that low load levels not induce structural failure.

Yanni [7] investigated the tensile creep of UHPFFC with different heat treatment. He included in his research work also pre-cracked specimens. After one year loading period he observed no tensile creep failure for a load level of 40 % of the tensile strength and for a period of 90 days with a load level of 60 % no failure occurred. Yanni stated a linear creep behaviour of UHPC up to a load of 60 % of the tensile strength. Specimen tested with 70 and 80 % tensile strength level crept to failure within few minutes.

Other studies concerning the tensile creep behaviour of UHPFRC were performed by Switek et al. [8], who correlated the tensile creep to micro-crack formation. Kamen et al. [9] investigated the early age UHPFRC creep and observed viscoelastic creep at stress strength levels higher than 50 and 63 % of the tensile strength. However, none of these investigated the tensile creep in correlation with cracked cross sections in detail.

2 Experimental Program

Many are the parameter that could be investigated for the tensile creep in cracked cross section. Certainly the most important parameter is the load level, together with the pre-damaging: the load or deformation applied to the specimen to generate the crack to be investigate. Regarding the steel fibre, the length/diameter aspect ratio (slenderness), the volume content, the fibre length, the surface property and the fibre orientation along the loading direction seems to influence the tensile creep. Concerning the concrete, the mix design and the presence of coarse aggregate may influence the creep, as well as heat treatment, which modifies the concrete structure and the cement to fibre adhesion. While for ordinary fibre reinforced concrete [5] the compressive strength seems to be decisive, the influence is less important for the UHPC [7].

In Table 1 the test program for the sustained load tests is summarized. There the fibre length, diameter and content, the load level and the adopted test setup are given.

Table 1 Sustained load test program

Series	Test description	Fibre length/Ø, content	Sustained load [% of tensile strength]	Test setup, number of tests	
No.				Axial	Bending
1	Reference combination	12.5/0.175, 2 %	25/40/80/90	2/4/2/2	2/4/2/2
2	Pre damaging effect	12.5/0.175, 2 %	40	2	2
3	Fibre aspect ratio	12.5/0.40, 2 %	40	2	2
4	Fibre content	12.5/0.175, 4 %	40/80	2/2	2/2
5	Without heat treatment	12.5/0.175, 2 %	40	2	2
6	Undamaged, creep test	12.5/0.175, 2 %	40/80	2/2	2/2
	Total loaded specimen			24	24

Table 2 UHFRC mix design

Material		
Cement CEM 52.5 N	728	kg/m^3
Water	80	kg/m^3
Sand DM 0.125/0.5	816	kg/m^3
Quartz flour	510	kg/m^3
Microsilica suspension	226	kg/m^3
Super plasticizer	29.7	kg/m^3
Steel fibre (2 %)	164	kg/m^3

A reference combination was tested extensively with a larger amount of specimen at four different load levels as a comparison with the other tests series. The mix design shown in Table 2 was employed, changing in some cases only the fibre diameter and content. Only the straight brass alloy coated fibres were used. This fibre type is mostly preferred and has a good interaction with the concrete paste [10]. For investigating the effect of the fibre content, four tests with 4 % fibre volume were performed in addition to the reference fibre concentration of 2 %. The slenderness of the fibre was investigated with a fibre volume of 2 % also and a length-diameter ratio of 31.3 (fibre length 12.5 mm, diameter 0.40 mm, reference ratio approx. 71 with 0.175 mm diameter).

For each test series six dog-bone specimens and six prisms were tested according to the French guideline [11] for the tensile strength and post cracking behaviour. With these tests, a deformation level was defined for the pre-damaging of the specimens in the sustained load tests (test series 1, 3, 4 and 5). This deformation was vertical displacement of 0.6 mm for the prisms and a axial deformation of 0.01 % for the dog-bone specimen. Since specimen with different fibres have different strength, the sustained load level was defined as a percentage of the respective reference tests. The pre-damaging effect was investigated with test series 2, with a larger pre-damaging deformation.

In order to exclude the superposition with basic and drying creep of the concrete paste, all the specimen were subjected to a heat treatment of 48 h at 90 °C (except test series 5). Heat treatment gives a very important reduction on the tensile and compressive creep [1, 7, 11, 12]. However since the effects of the uncracked concrete creep cannot be neglected, in test series 6 four uncracked specimen were included in the test program.

Regarding the test setup adopted for this investigation, four point bending and axial tensile tests were chosen. The four point bending test allow to observe the average deformation of a large portion of the material compared to three point bending test with notched specimen. Four point bending tests were performed on prisms 70 × 70 mm according to the French guideline [11] but with a span of 940 mm between the supports, instead of 210 mm. The span was increased to limit the required dead load and to increase the observed surface. The vertical displacement of the midpoint was recorded. A sustained load frame with lever arms for axial tensile tests was built. This allows to test up to 24 specimen simultaneously.

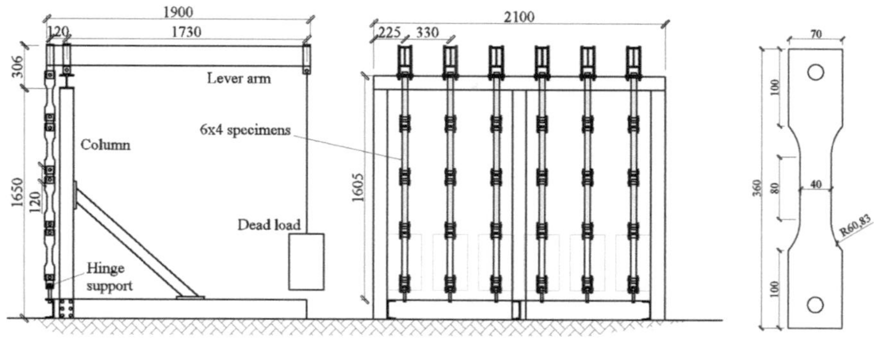

Fig. 1 Sustained load test setup and dog-bone specimen

Dog-bone specimen were designed as shown in Fig. 3. The forms for these specimens were CNC milled in order to guarantee a very precise geometry and avoid unwanted eccentricity (Fig. 1).

One of the decisive issue for the test was the choice of the measuring method. Since there are a lot of specimen that have to be monitored over long time. Beside conventional mechanical displacement indicators, 48 potentiometer displacement transducers were installed on the dog-bone specimen.

Since the expected deformations are quite small, together with the displacement transducers also survey points for extensimeter measurement were fixed at the dog-bone specimen.

3 UHPFRC Material Behaviour

The concrete mix used for this investigation is indicated in the Table 2. This concrete has a compression resistance between 150 and 190 Mpa. The steel fibres are manufactured by the company Krampeharex and are straight brass alloy coated steel wires. The water-cement ratio, under consideration of the water in the silica suspension, is 0.265.

In a compression test, the material has a quite elastic linear behaviour up to failure. The addition of fibre do not influence the load peak. The most common tests to characterize the tensile strength and the post cracking behaviour are the three and four point bending tests and axial tension tests with notched and unnotched specimens [11]. The difference between the three and four point bending tests is presence of a notch in the three point test specimens, which fixes the point where the crack appears. Having only one crack in a defined position allows easier monitoring the width over the time.

Figure 2 shows reference tests on the adopted concrete with three different setups, all performed according the AFGC French guideline for UHPFRC [11]. In general, the diagram starts with a linear elastic part, where the uncracked concrete matrix resists to the tensile forces, almost without influence of the fibres. With the formation of the first crack the tensile load is transferred from the concrete to the

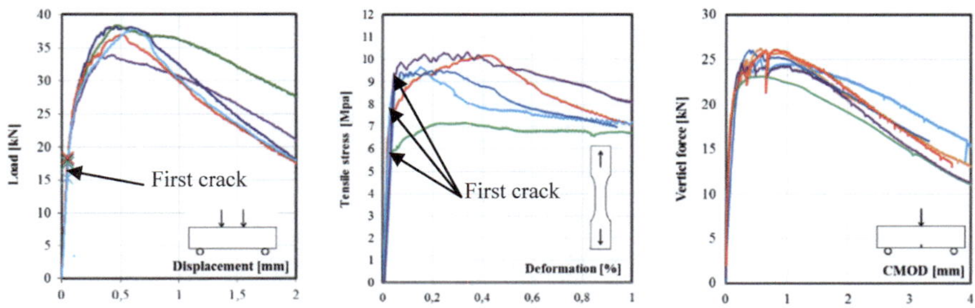

Fig. 2 Reference tests, from *left*, 3 point bending, four point bending and centric tension tests

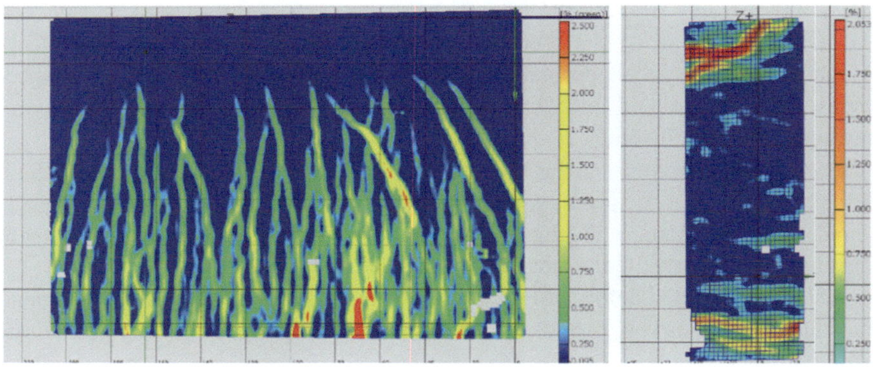

Fig. 3 Aramis image analysis, four point bending (*left*) and axial tensile test (*right*)

fibres that have to bridge the tensile forces over the cracked cross section. After the formation of the first crack, the applied load can be further increased. The tensile resistance of the concrete is exceeded then in several cross sections. The results is a multiple fine crack scenario difficult to observe with naked eye. In general, after the peak load, the failure occurs with crack localization: only one crack becomes larger and larger and the specimen slowly collapses.

To investigate in detail the evolution of the damage and the cracks progression in the cross section the optical measurement system ARAMIS was adopted. Figure 3 shows the analysis of the specimens surface with this system. The bending test, in comparison with the axial tensile test, shows a larger increase of tension load after the first crack up to failure and a larger number of cracks. These effects are caused by the internal force redistribution in the cross section and by the progressive activation of the fibres during the crack opening process. In the three point bending test, instead of the vertical displacement, the crack mouth opening displacement (CMOD) is measured with a clip gage. In this test only one crack appears. This observation allows to correlate the deformation of the cracked cross section with the internal stresses.

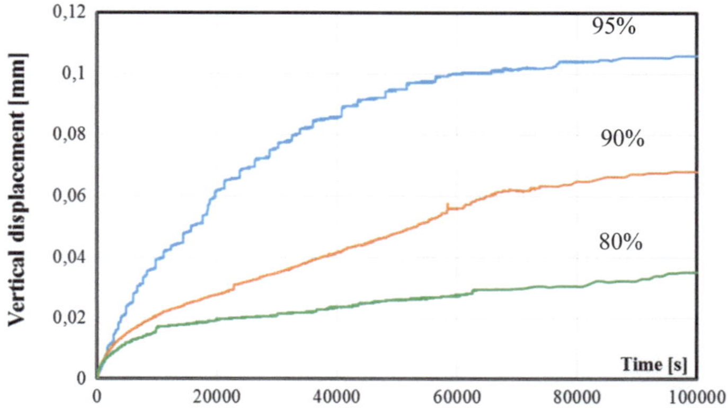

Fig. 4 Preliminary tests, vertical displacement on four point bending test

3.1 Preliminary Sustained Load Tests

Preliminary tests in a universal testing machine were performed in order to investigate the sensitivity of the material to tensile creep. A first result is the order of magnitude of the displacements. Specimens for the four point bending test according to [11] with a cross section of 70 × 70 mm and span of 210 mm were subjected to sustained load over 27 h. For the pre-damaging, the specimen were first loaded up to a deflection of 0.6 mm, as described above. This procedure provide the specimen the required crack pattern. Specimens were then unloaded and reloaded up to 95, 90 and 80 % of the load reached with the imposed deflection. In Fig. 4 the vertical displacement during the sustained load is shown. Since the time of observation was only few hours, a stabilization of the vertical displacement was not observed. Higher applied load results in level higher level of displacements. The tensile creep seems to be non-linear for such high load levels.

4 Fibre Orientation and Computed Tomography

For the investigation of a correlation between the experimental results and the fibre distribution, some of the specimen will be analysed with the computed tomography (CT). This technique, together with the software MAVI ([13], Modular Algorithms for Volume Images) allows to measure directly the fibres concentration and their orientation. Particular interesting is to analyse the failure cross section and in the neighbourhood areas.

The computed tomography has demonstrated its potentiality for ordinary and for ultra-high performance fibre reinforced concrete as reported in [14]. Since it is quite hard to correlate the presence of fibre in the specimen and the tensile creep in cracked cross section, it was decided to investigate the orientation after the

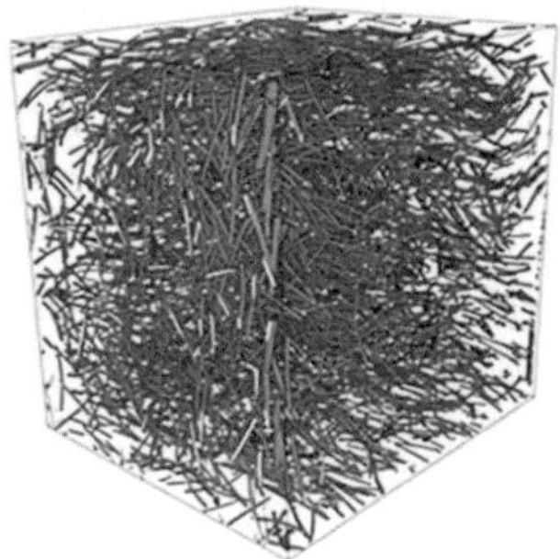

Fig. 5 Fibre 3d rendering reconstruction (fibre length 6 mm, Ø 0.175 mm), Schnell [14]

sustained load tests. Considerations about the fibre orientation and the post crack tensile behaviour are well described by Frettlöhr [15]. The biggest difference between ordinary fibre reinforced concrete and UHPC is the larger amount of fibres. This complicates the analyses of the single fibre. Moreover the dense structure of the concrete limits the size of the cross sections up to a dimension of about 40 mm. This is the reason why the axial tension specimens are limited in their cross section (Fig. 1). Figure 5 shows how a 3D reconstruction of the fibres in a 50 × 50 mm specimen separated from the concrete. With this reconstruction is possible to divide the volume in different areas and calculate the fibre concentration and the orientation along the three principal directions.

5 Conclusions

The long term tensile creep of cracked UHPFRC is object of a study at the TU Kaiserslautern and after a preliminary phase, sustained load tests on different specimens are initiated.

First tests with short term high sustained load showed a quite good response of UHPFRC even if they may be not representative for the long term behaviour of the material. Variable load has also an important influence on the creep and in combination with sustained load may play a decisive role for the safety of a structure.

The computer tomography for UHPC, even if burdensome and expansive, is a suitable method for studying the average fibre orientation and distribution, almost like for ordinary concrete. In case of small cross section, like that investigated in the tensile tests, it is possible to "scan" the whole specimen.

Further investigation have to consider other fibre surface, and the influence of different mix designs, including those with coarse aggregate. The tests with a lower load level have to be observed during the next several years.

The results of this investigation will deliver a decisive contribute to the sustainable construction of UHFRPC structural components, clarifying the reliability of the fibres even without conventional reinforcement bars.

Acknowledgments This research was possible thanks to Deutsche Forschungsgemeinschaft (DFG) that financed and implemented the implement a new research-training group (RTG) "Stochastic Models for Innovations in the Engineering Sciences" at the University of Kaiserslautern. Moreover, the acquisition of CT data is possible thanks to the cooperation with the Fraunhofer ITWM of Kaiserslautern.

References

1. Deutscher Ausschuss für Stahlbeton Ultrahochfester Beton—Sachstandsbericht, 1. Aufl. DAfStb Heft 561, Berlin (2008)
2. Candido, L., Micelli, F., Aiello, M. A., Plizzari, G.: Cracking behaviour of FRC beams under long-term loading. In: Hellmich, C., Pichler, B., Kollegger, J. (eds.) CONCREEP 10: Mechanics and Physics of Creep, Shrinkage, and Durability of Concrete and Concrete Structures Proceedings of the 10th International Conference on Creep, Shrinkage, and Durability of Concrete and Concrete Structures, September 21–23, 2015 Vienna, Austria, pp. 1147–1156. Reston, Virginia, American Society of Civil Engineers (2015)
3. Vasanelli, E., Micelli, F., Aiello, M. A., Plizzari, G.: Long term behaviour of fiber reinforced concrete beams in bending. In: Barros, J. A. (ed.) Eighth RILEM International Symposium on Fibre Reinforced Concrete (BEFIB 2012) (Guimarães, Bagneux, Univ. do Minho; RILEM Publ, 2012 RILEM proceedings PRO, 88) (2012)
4. Zhao, G., Di Prisco, M., Vandewalle, L.: Experimental investigation on uniaxial tensile creep behavior of cracked steel fiber reinforced concrete. Mater. Struct. **48**(10), 3173–3185 (2015)
5. Arango, S., Taengua, E. G., Vargas, J. R., Serna Ros, P.: A comprehensive study on the effect of fibres and loading on flexural creep of SFRC, BEFIB (2012)
6. Bernard, E. S.: Influence of fiber type on creep deformation of cracked fiber-reinforced shotcrete panels. Mater. J. **107**(5) (2010)
7. Yanni, V. Y. G.: Multi-scale investigation of the tensile creep of ultra-high performance concrete for bridge applications. Ph.D thesis, Georgia Institute of Technology, Georgia Institute of Technology, December (2009)
8. Switek, A., Denarié, E., Brühwiler, E.: Modeling of viscoelastic properties of ultra high performance fiber reinforced concrete (UHPFRC) under low to high tensile stresses. In: ConMod 2010: Symposium on Concrete Modelling (2010)
9. Kamen, A., Denarié, E., Sadouki, H., Brühwiler, E.: UHPFRC tensile creep at early age. Mater. Struct. **42**(1), 113–122 (2009)
10. Corinaldesi, V., Nardinocchi, A., Donnini, J.: The influence of expansive agent on the performance of fibre reinforced cement-based composites. Construct. Build. Mater. **91**, 171–179 (2015)
11. Association Française de Gènie Civil (AFGC): Documents scientifique et technique: Bétons fibrés à ultra-hautes performances, Edition révisé, Juin 2013 (Recommandations) (2013)
12. Garas, V. Y., Kurtis, K. E., Kahn, L. F.: Creep of UHPC in tension and compression. Cem. Concr. Composites **34**(4), 493–502 (2012)
13. Fraunhofer Institut für Techno-und Wirtschaftsmathematik, MAVI-Modular Algorithms for Volume Images
14. Schnell, J., Schladitz, K., schuler, F.: Richtungsanalyse von Fasern in Betonen auf Basis der Computer-Tomographie, Beton-und Stahlbetonbau **105**(2), 72–77 (2010)
15. Frettlöhr, B.: Bemessung von Bauteilen aus ultrahochfestem Faserfeinkornbeton, Stuttgart, Universität Stuttgart, Institut für Leichtbau Entwerfen und Konstruieren, (Ph.D. Thesis, 9 Dec 2011

Experimental Study on Time-Dependent Behavior of Cracked UHP-FRCC Under Sustained Loads

Tomoya Nishiwaki, Sukmin Kwon, Hiroto Otaki, Go Igarashi, Faiz U.A. Shaikh and Alessandro P. Fantilli

Abstract Ultra-high-performance fiber-reinforced cementitious composite (UHP-FRCC) has been successfully developed over the last decade. UHP-FRCC shows very high strength and high durability because of its very low water binder ratio and dense microstructures. Moreover, by incorporating a combination of steel and mineral fibers with different sizes in the mixture design, UHP-FRCC also exhibits high ductility and high energy absorption capacity, even under uniaxial tensile stress. The high ductility of UHP-FRCC materials shows strain hardening behavior with multiple cracking after the occurrence of the first crack. While extensive research on the tensile and bending behavior of this composite under short-term loading is reported, only limited results can be found on time-dependent behavior of UHP-FRCC under sustained long-term loading, which is strongly related to the situation of real structures. This paper presents experimental results concerning the time-dependent behavior of cracked UHP-FRCC specimens under sustained bending load. In bending creep test, sustained load with 50 % of the pre-cracking load was applied to a pair of prismatic specimens of 50 × 100 × 400 mm under a 4-point bending setup. The performance of UHP-FRCC is compared with those of conventional fiber reinforced cementitious composites (FRCC) using steel and synthetic fibers, which are investigated within the Round Robin Test (RRT) organized by RILEM TC 261-CCF. The results showed that UHP-FRCC exhibited smaller creep deformation at the sustained loading condition than the conventional FRCC, even when the initial deformation of the UHP-FRCC is much

T. Nishiwaki (✉) · S. Kwon · H. Otaki · G. Igarashi
Department of Architecture and Building Science,
Graduate School of Engineering, Tohoku University, Sendai, Japan
e-mail: tomoya.nishiwaki.e8@tohoku.ac.jp

S. Kwon
The ACE-MRL, University of Michigan, Ann Arbor, USA

F.U.A. Shaikh
Department of Civil Engineering, Curtin University, Perth, Australia

A.P. Fantilli
Department of Structural, Building and Geotechnical Engineering,
Politecnico di Torino, Turin, Italy

P. Serna et al. (eds.), *Creep Behaviour in Cracked Sections of Fibre Reinforced Concrete*, RILEM Bookseries 14,
DOI 10.1007/978-94-024-1001-3_9

larger than the conventional FRCC. Cracking pattern, in terms of crack width and number of cracks, was also studied in both composites under sustained loads. As a result, the superior resistance of UHP-FRCC was obtained with respect to conventional FRCC.

Keywords Ultra-high-performance fiber-reinforced cementitious composite (UHP-FRCC) · Bending load · Creep · Multiple cracks · Crack width

1 Introduction

UHP-FRCC is categorized as strain hardening cementitious composites (SHCC) and shows excellent mechanical properties even under tensile stress. In particular, strain hardening behavior has been achieved with multiple cracking instead of widening of a single crack. The number of cracks increases during strain hardening but the widths of multiple cracks can be controlled within very narrow range, generally less than 0.1 mm. Due to the very dense microstructure with the extremely low water-binder ratio (0.15 in this study), UHP-FRCC shows excellent performance against penetration of aggressive substances into the matrices [1]. On the other hand, although the crack widths are very small the presence of multiple cracks is mandatory for expected superior mechanical performance. In other words, the UHP-FRCC should be designed in cracked state, if their excellent mechanical properties have to be taken into account. Thus, also the durability of UHP-FRCC should be discussed in cracked state, even the crack width of each generated multiple-crack is small enough.

In this study, time-dependent behavior up to 28 days of UHP-FRCC was experimentally evaluated. Generally, the cracked conventional fiber reinforced cementitious composites (FRCC) and/or cracked SHCC under sustained load show deformation due to creep with widening of existing crack width (e.g. [2]). However, only few results of creep and shrinkage behavior of UHP-FRCC are available in the current literatures. Bending creep tests have been carried out in order to evaluate the time-dependent behavior of UHP-FRCC under sustained load. In addition, the performance of UHP-FRCC is compared with those of conventional FRCC containing steel and synthetic fibers, which are the object of the Round Robin Test (RRT) organized by RILEM TC 261-CCF.

2 Employed UHP-FRCC in This Study

The UHP-FRCC employed in this study is based on the previous studies conducted by the authors [3, 4]. In order to improve the tensile performance of UHP-FRCC, the factors that influence the tensile behavior of UHP-FRCC are clarified and a material design concept based on a multi-scale fiber-reinforcement system

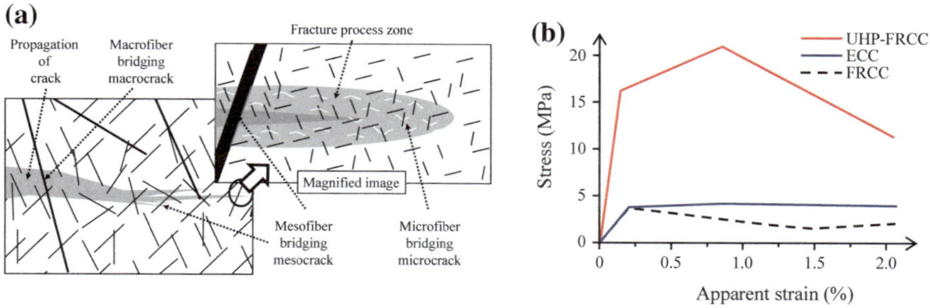

Fig. 1 Employed UHP-FRCC; **a** schematic illustration of multi-scale fiber-reinforcement system [4], **b** schematic diagram of tensile behavior

(i.e. each scale of fibers reinforces each scale of cracks) is proposed and shown in Fig. 1a. According to fracture mechanics criterion the cracks in concrete initiate at the micro-scale as lots of very fine and invisible cracks are formed where tensile stress is concentrated. With further increase in load, these cracks unify and propagate into the concrete matrix. This process progresses to the meso-scale and macro-scale, which finally leads to failure. Fibers with proper size can bridge cracks and restrain its widening. UHP-FRCC can be obtained by reinforcing a very high strength matrix with multi-scale fibers. Typical mechanical performance of UHP-FRCC employed in this study is shown in Fig. 1(b), which describes the relationship between strain and tensile stress during the uniaxial tensile test. Strain hardening behavior and high tensile strength that is greater than 20 MPa are observed in UHP-FRCC in comparison to the conventional FRCC.

3 Experimental Program

Bending creep experiments with cracked UHP-FRCC specimens was carried out in order to measure the time-dependent behavior and to compare the performance with conventional FRCC of the ongoing RRT organized by TC 261-CCF. Initial cracks were introduced by pre-bending loading. Employed materials and mix proportion of UHP-FRCC are shown in Tables 1 and 2, respectively. The mix proportions of the RRT specimens are also shown in Table 2. For the RRT, 2 types of FRCC containing 2 different fibers, which are hook-ended steel fiber (fiber length is 60 mm, hereafter FRCC-ST series) and synthetic fiber (fiber length is 40 mm, hereafter FRCC-SY series) were employed. The geometry of employed specimens and test set-up for sustained loading are shown in Fig. 2. Note that the geometries of UHP-FRCC (prism shape with 50 × 100 × 400 mm) and FRCC-ST and FRCC-SY specimens (prism shape with 150 × 150 × 600 mm) are different from each other due to the capacity of the creep test set-up. The numbers in brackets in Fig. 2 show the size of the FRCC-ST and FRCC-SY specimens and its test set-up.

Table 1 Employed materials for UHP-FRCC specimens

Materials	Symbol	Density (g/cm^3)	Properties
Cement	C	3.01	Low-heat cement and silica fume were premixed at mass fraction of 82 and 18 % (Blaine fineness: 6555 cm^2/g)
Fine aggregate	S	2.6	Natural silica sand (average particle size: 0.212 mm)
Superplasticizer	SP	1.05	Polycarboxylate acid system
Anti-foaming agent	D	1.05	Polyether system
Wollastonite	Wo	2.9	Length: 50–2000 μm, aspect ratio: 3–20, tensile strength: 2700–4100 N/mm^2, Young's modulus: 303–530 kN/mm^2
Short steel fiber	SS	7.85	Straight shape, length: 6 mm, diameter: 0.16 mm, aspect ratio: 37.5, tensile strength: 2000 N/mm^2, Young's modulus: 206 kN/mm^2
Long steel fiber	LS	7.85	Hooked end shape, length: 30 mm, diameter: 0.38 mm, aspect ratio: 78.9, tensile strength: 3000 N/mm^2, Young's modulus: 206 kN/mm^2

Table 2 Mix proportion of UHP-FRCC and FRCC of the RRT

Series	C[a] (kg/m^3)	W (kg/m^3)	SP (kg/m^3)	D (kg/m^3)	S[b] (kg/m^3)	G[c] (kg/m^3)	Wo (kg/m^3)	Fiber (vol.%)	Compressive strength (MPa)
UHP-FRCC (7)	1513	205	21	0.3	530	–	167	SS1.5 + LS2.0	187.5
UHP-FRCC (28)									204.1
FRCC-ST	350	175	–	–	1132	709	–	0.38	34.8
FRCC-SY								0.10	35.7

[a]Cement for FRCC-ST and FRCC-SY series is CEM I 42.5R

[b]Fine aggregate for FRCC-ST and FRCC-SY series are river sand

[c]*G* gravel aggregate, limestone

In addition, the UHP-FRCC specimens do not have a notch since multiple cracks are expected in the constant moment zone. The specimens of the FRCCs were provided by the TC 261-CCF. The UHP-FRCC specimens were demolded after 24 h from casting, and then steam cured at 90 °C for 24 h. Afterward, the UHP-FRCC specimens were settled in a climate room (20 °C and 40 %RH) for 7 days [named UHP-FRCC(7)] and 28 days [named UHP-FRCC(28)]. As shown in Table 2, the compressive strength of UHP-FRCC is much higher than the conventional concrete and UHP-FRCC(28) shows 10 % higher compressive strength than the UHP-FRCC(7).

Four-point bending test was carried out to generate initial crack by a universal testing machine (capacity: 100 kN). The deformation measured by pi-gauge was

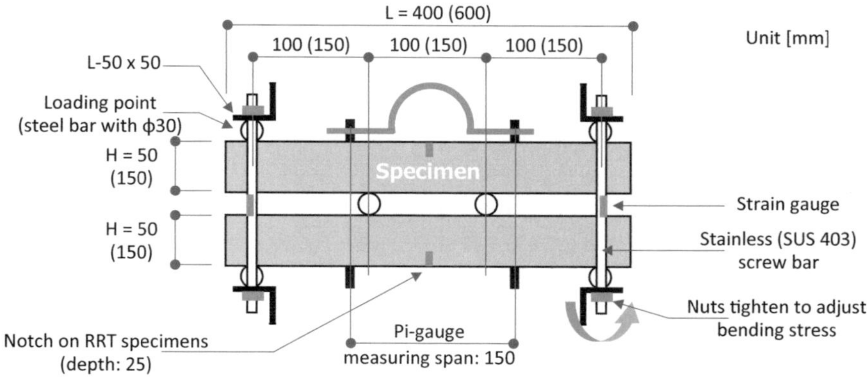

Fig. 2 Specimen and set-up for sustained loading

introduced up to 1.0 mm for UHP-FRCC specimens and up to 0.5 mm for the FRCC-ST and FRCC-SY specimens. After the bending test, cracked specimens were placed on a creep test set-up shown in Fig. 2 to apply sustained loading. The sustained bending load was adjusted to 50 or 60 % of the pre-cracking loading by rotating nuts at the edge of screw bar columns, and load value was calculated from the strain of the screw bar columns (SUS 304, Young's modulus: 193 kN/mm^2). During bending creep test, deformations at the tension side (bottom side) of each specimen were measured by pi-gauges. The bending creep tests were conducted in the climate room. At the same time, other prismatic specimens with 100 × 100 400 mm were exposed to the same condition in order to measure dry shrinkage. Regression curve of dry shrinkage was drawn according to the ACI committee report [5] as the following equation;

$$\varepsilon_{sh,t} = a \cdot \left[\frac{t}{35 + t}\right] \quad (1)$$

where, $\varepsilon_{sh,t}$ is dry shrinkage and t is time. a is empirical constant (= 168.0 for UHP-FRCC, 246.0 for FRCC-SY and 157.0 for FRCC-ST). The deformation due to creep was calculated omitting the shrinkage phenomena.

4 Results and Discussion

Figure 3 shows the bending stress vs deflection for UHP-FRCC and bending stress vs CMOD for FRCC-ST and FRCC-SY specimens. It should be noted that the UHP-FRCC specimens had no notch because of their multiple-cracking behavior, therefore not CMOD of a single crack but deformation of the bottom of the bending span measured by pi-gauge was employed. Regarding the pre-cracking load, both

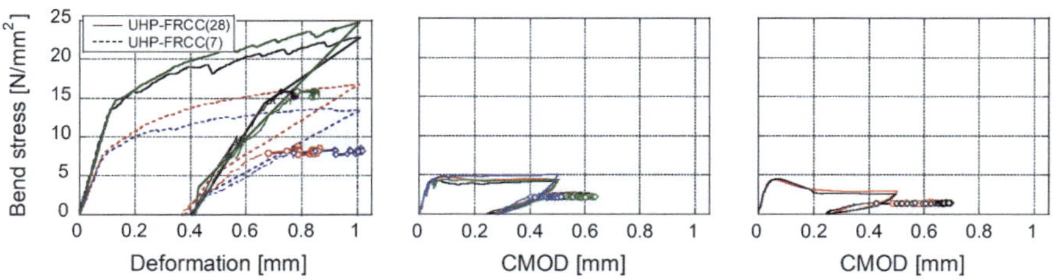

(a) UHP-FRCC series (b) FRCC-ST series (c) FRCC-SY series

Fig. 3 Relationship between deformation/CMOD and bending stress during pre-cracking loading to creep loading up to 28 days

UHP-FRCC series showed deflection hardening behavior and much higher bending stress than the conventional FRCCs with the maximum bending stress of 5 MPa. Moreover, the ultimate stresses of UHP-FRCC(7) and UHP-FRCC(28) were higher than the released stress of 15 and 24 MPa respectively since the load released point was in the middle of deflection hardening. Because of the difference in the geometry of specimens, deformations cannot be directly compared to each other. In order to compare the deformation and CMOD, the average curvatures κ were calculated from these values of the displacement by means of the following Eq. (2);

$$\kappa = \frac{\Delta x}{{}^{h}\!/_{2} \times L} \tag{2}$$

where, h is height of the specimen (= 50 mm for UHP-FRCC and 150 mm for FRCC), L is length of constant moment zone of the spesimen (= 100 mm for UHP-FRCC and 150 mm for FRCC), and Δx is displacement measured by pi-gauge (deformation for UHP-FRCC and CMOD for FRCC). In the pre-cracking test, deformation of 1.0 mm on UHP-FRCC specimens and CMOD of 0.5 mm on FRCC specimens are converted to 1/2500 and 1/22,500, respectively. That is, the initial displacement of the UHP-FRCC specimens can be recognized as 9 times the deformation of the FRCC specimen. Although the curvature of the UHP-FRCC specimens was superior to the FRCC specimens, creep deformation on the UHP-FRCC(7) and UHP-FRCC(28) specimens did not exceed the initial deformation of 1.0 mm, contrary to creep CMOD on the FRCC specimens which exceeded the initial 0.5 mm CMOD, as shown in Fig. 3.

Creep coefficient φ was calculated by the following Eq. (3);

$$\varphi = \frac{\varepsilon_t - \varepsilon_{to}}{\varepsilon_{to}} \tag{3}$$

where, ε_t is displacement per unit bending stress including dry shrinkage at the time t, and ε_{t0} is displacement per unit bending stress including dry shrinkage at the time

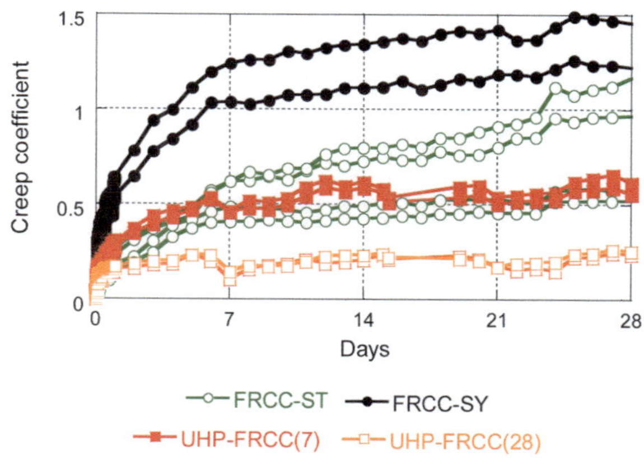

Fig. 4 Relationship between loading time and creep coefficient up to 28 days

to start to subject sustained bending load t_0. Here, ε_t and ε_{t0} are obtained from the following equations;

$$\varepsilon_t = \frac{D_t - D_{sh,t}}{\sigma_t} \tag{4}$$

$$\varepsilon_{t0} = \frac{D_{t0} - D_{sh,t0}}{\sigma_{t0}} \tag{5}$$

where, D_t and D_{t0} are the measured displacements by pi-gauges at the time t and t_0, $D_{sh,t}$ and $D_{sh,t0}$ are displacements due to shrinkage at the time t and t_0, σ_t and σ_{t0} are bending stress at the time t and t_0, respectively. $D_{sh,t}$ and $D_{sh,t0}$ were calculated according to Eq. (1). In this experimental campaign, since the bending stress was slightly lost and was needed to be adjusted via screw bolts and nuts during sustained loading, obtained values of displacements were divided by real-time bending stress. The obtained creep coefficients are shown in Fig. 4 as the relationship with loading time. This graph clearly shows that the creep coefficient of UHP-FRCC series was much lower than the FRCC in both ST and SY series, although the curvature of UHP-FRCC series was much higher than the FRCC specimens. In particular, in the UHP-FRCC(28) the creep coefficient increased only in the first day and showed almost the same value after the second day with no significant increase in creep coefficient with progress of time until 28 days. Almost similar behaviour was also observed in the case of UHP-FRCC(7), but with slightly higher creep coefficient values. On the other hand, in case of FRCC-ST and FRCC-SY series different behaviour is observed. The creep coefficients increased rapidly up to 7 days and the rate of increase of creep coefficient was slowed down after 7 days and increased at lower rate until 28 days. It is also observed that the FRCC-SY series exhibited much higher creep coefficient than that of FRCC-ST series.

Figure 5 shows the crack pattern on the surface of the UHP-FRCC(7) and UHP-FRCC(28) specimens before [(a1) and (b1)] and after [(a2) and (b2)] 28 days creep loading, respectively. Red lines in Fig. 5a2, b2 indicate the cracks generated

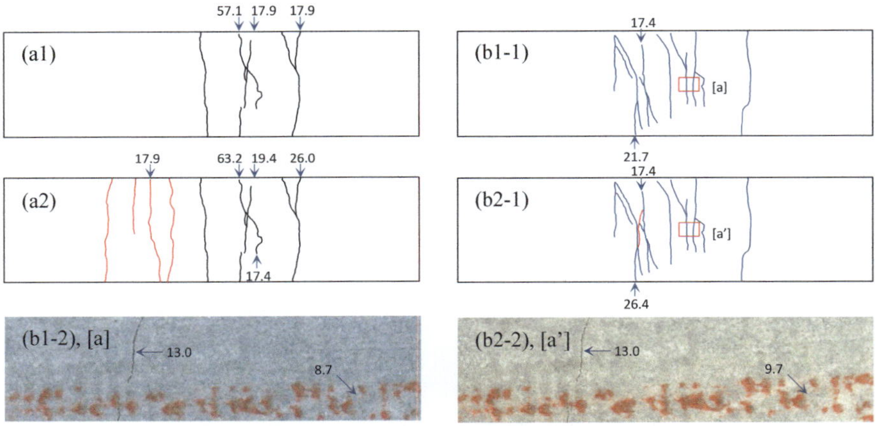

Fig. 5 Crack formation on the surface of a specimen; **a1** UHP-FRCC(7) at the beginning of sustained loading, **a2** UHP-FRCC(7) after sustained loading, **b1-1** UHP-FRCC(28) at the beginning of sustained loading, **b2-1** UHP-FRCC(28) after sustained loading, **b1-2** magnified area of (*a*) and **b2-2** magnified area of (*a*′)

under sustained loading. Numbers next to the cracks show its width measured with a micrometer scale (note that the cracks without number have a width below 15 μm). In the case of the FRCC specimens, displacement due to creep load appears as an increment of CMOD of single crack as shown in Fig. 3, since a notch is installed on the specimen. On the other hand, the numbers of cracks are increased in UHP-FRCC series. However, the crack widths were not significantly changed, as shown in Fig. 5b believed to be due to deflection (strain) hardening and multiple cracking characteristics of this material, which occurred under static bending (tensile) loading.

In this paper, only the results up to 28 days are shown as an intermediate report of this test. Since the other tests are still undergoing, the results with longer creep period will be provided in future reports.

5 Conclusions

- UHP-FRCC with multi-scale fiber-reinforcement system showed hardening behavior under 4-point bend test. The bending strength was much higher than the conventional FRCC employed in the RRT organized by RILEM TC 261-CCF.
- Creep deformation and creep coefficient of UHP-FRCC up to 28 days was much smaller than the conventional FRCCs, although the average curvature of UHP-FRCC specimen was 9 times than that of FRCCs. In particular, the creep coefficient of UHP-FRCC(28) series was below 0.3.

– During creep loading, deformation of the UHP-FRCC was a consequence of the generation new narrow cracks, as well as the slightly widening of the initial cracks. In other words, the creep deformation of UHP-FRCC can be recognized as hardening behavior under static sustained loading.

Acknowledgments This work was partially supported by JSPS KAKENHI Grant Number 25630228, 26289186 and 26-7167. The authors would like to express their deep gratitude.

References

1. Tanaka, Y., Musha, H., Tanaka, S., Ishida, M.: Durability performance of UFC sakata-mira footbridge under sea environment, FraMCoS-7, pp. 1648–1654 (2010)
2. Boshoff, W.P.: Cracking behavior of strain-hardening cement-based composites subjected to sustained tensile loading. ACI Mater. J. **111**, 553–559 (2014)
3. Kwon, S., Nishiwaki, T., Kikuta, T., Mihashi, H.: Development of ultra-high-performance hybrid fiber-reinforced cement-based composites. ACI Mater. J. **111** (2014)
4. Kwon, S., Nishiwaki, T., Choi, H., Mihashi, H.: Effect of wollastonite microfiber on ultra-high-performance fiber-reinforced cement-based composites based on application of multi-scale fiber-reinforcement system. J. Adv. Concr. Technol. **13**, 332–344 (2015)
5. ACI Committee 209: Prediction of creep, shrinkage, and temperature effects in concrete structures. American Concrete Institute (1997)

Creep Behaviour of Cracked High Performance Fibre Reinforced Concrete Beams Under Flexural Load

Eduardo Galeote, Ana Blanco, Albert de la Fuente and Sergio H.P. Cavalaro

Abstract The investigation on flexural creep of high performance fibre reinforced concrete (HPFRC) is still scarce. Even though the presence of fibres in concrete helps to control the deformations, these may increase under the effect of a sustained load. To analyse the effect of creep in pre-cracked HPFRC elements, twelve beams reinforced with either glass or steel fibres with dimensions 40 × 80 × 1200 mm were tested under a three-point configuration. For that, a new type of frame was designed and constructed to test the HPFRC beams under flexural load in a climate-controlled room with constant temperature and relative humidity. The loading mechanism was based on a lever system, applying sustained load ranging between 25 and 50 % of the load at which the first crack appeared. The deflection at the mid-span was registered by means of LVDT transducers. Additionally, the influence of the curing procedure (with or without aluminium tape wrap) was assessed. In general, glass fibre reinforced beams presented higher deflections than steel fibres, even though at low load levels the type of fibre did not have significant influence on the deformation.

Keywords HPFRC · Creep · Steel fibres · Glass fibres · Flexural load

1 Introduction

High performance fibre reinforced concrete (HPFRC) represents an important innovation in the field of construction materials with a high potential of application [1]. The inherent brittleness of the matrix of this type of concrete may be partly compensated by the addition of fibres, which allows increasing ductility at the cracked state depending on the type and content of fibres used.

E. Galeote (✉) · A. Blanco · A. de la Fuente · S.H.P. Cavalaro
Department of Civil and Environmental Engineering,
Universitat Politècnica de Catalunya, Barcelona, Spain
e-mail: eduardo.galeote@upc.edu

P. Serna et al. (eds.), *Creep Behaviour in Cracked Sections of Fibre Reinforced Concrete*, RILEM Bookseries 14,
DOI 10.1007/978-94-024-1001-3_10

The effect of creep in concrete is of paramount importance and should not be ignored in design since extreme deformations can compromise serviceability [2]. Unfortunately, there is not a specific criterion to evaluate this effect neither in guidelines nor in codes. The lack of a unified methodology hinders the analysis of results already published due to the significant differences on the methods used. Besides, the long time required to evaluate creep makes its research less common due to the difficulty in obtaining results. However, some studies about creep may be found in both cracked and non-cracked sections in fibre reinforced concrete (FRC) [3–5].

The main aim of this research was to develop a new method capable of measuring the deformation produced by creep in slender elements. For this reason, a steel frame able to hold up to 12 beams was designed and constructed. This structure is based on a system of levers to transmit the load to the mid-span of the beams and includes an acquisition data system to measure the deflection at the central section of the beam. The primary advantage is that this structure enables to load and measure the deformation of each beam individually. Such a type of frame allows to independently customize the load to test each beam without interfering on the others.

An experimental program was conducted to analyse both the performance of the structure and the creep in HPFRC beams. The analysis of creep involved the manufacturing and testing of 12 beams with a height of 40 mm, a width of 80 mm and a length of 1200 mm with two different types of fibre reinforcement and two different curing processes in a pre-cracked state. Additionally, and as a result of the post-peak behaviour of the beams during the pre-cracking process, different load levels were applied onto the beams.

2 Design of the Frame

One of the main goals pursued in the design of the new frame was that each beam could be loaded individually. Unlike other test setups previously proposed [6, 7] in which a column for three beams was designed, the solution here presented required more space since the beams were not piled but placed one next to another. Some initial estimations displayed a strength of the concrete at the mid-span high enough to think about using a system of levers to transmit and easily handle the loads. Moreover, the creep was evaluated by means of the deflection instead of measuring the crack opening or both parameters [5, 8].

The steel frame (Fig. 1) was designed to measure the creep of 12 beams at the same time. The frame consists of a system based on a three-point configuration with a distance between supports of 1100 mm (Fig. 2b). A hollow steel cylinder was placed at the centre of the upper face of each beam together with a steel sheet. Fixed to the frame, an aluminium structure above the levers and the beams is used to hold the LVDTs to measure the deflection of the beams. These LVDTs are in direct

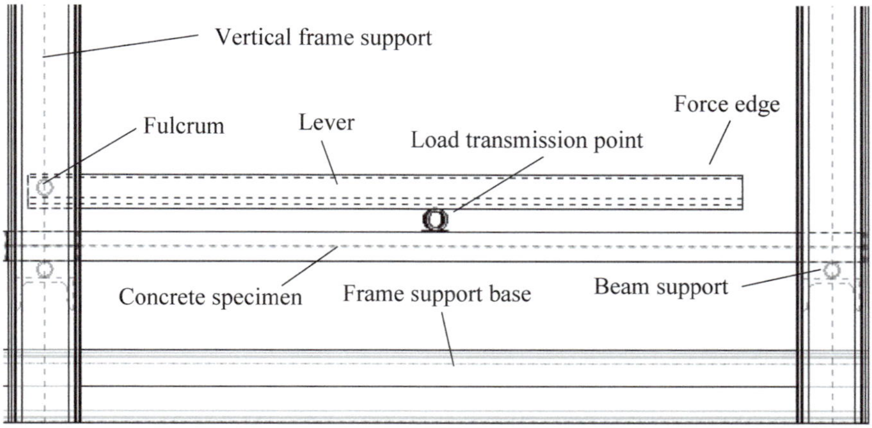

Fig. 1 Schematic representation of the frame for the creep test

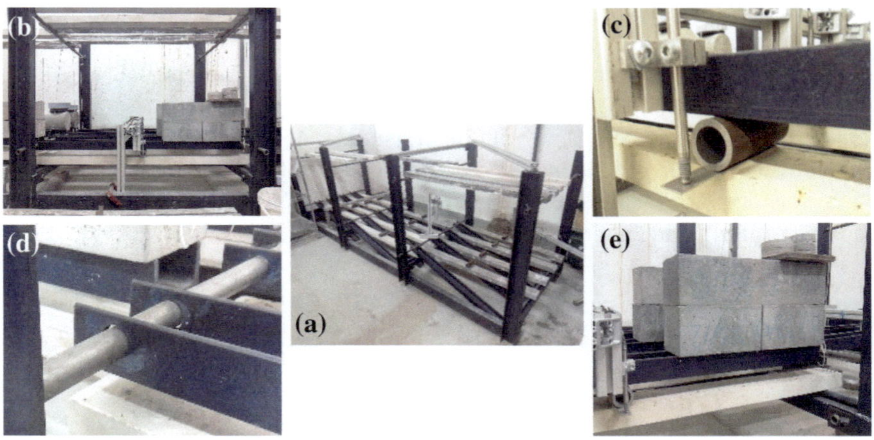

Fig. 2 **a** Steel frame, **b** three-point configuration 1100 mm length, **c** hollow cylinder with LVDT transducer, **d** hinge mechanism and **e** load on lever

contact with the steel sheets in the centre of the beams (Fig. 2c) and connected to a data acquisition system.

At the vertical columns, hollow steel bars of 22 mm of diameter were installed crosswise to produce the effect of a hinge mechanism to the levers as depicted in Fig. 2d. The levers were fixed to these bars at one edge and were free at the opposite edge, allowing the rotation around the longitudinal axis of the steel bar. To transmit the load to the beams, the levers rest on the steel cylinders fixed to the beams while the loads are placed on the free edge of the steel beam as shown in Fig. 2e.

3 Experimental Program

3.1 Materials and Concrete Mix

Two series of concrete were produced following the same mixing procedure. The type of fibre, after which the mixes were named, was the main difference between the two series manufactured. SF contained 150 kg/m^3 of steel fibres, whereas GF had 44 kg/m^3 of glass fibres. These two contents of fibres represent a volume of 1.9 % in SF and 1.6 % in GF with respect to the total concrete volume. Further details about the proportioning of the mixes and the properties of the fibres are presented in Tables 1 and 2, respectively.

The characterization of the material involved the elastic modulus (E_{cm}), which was determined according with [9] in 4 cylindrical specimens of $\phi 150 \times 300$ mm. The compressive strength (f_{cm}) was tested according with [10] in 4 cubic specimens of 150×150 mm and the residual tensile strength (f_{R1},f_{R2},$f_{R3}$and f_{R4}) was assessed by means of the three-point bending test in specimens of 150×150 600 mm according with [11]. Table 3 presents the average results of these tests together with the coefficient of variation (CV).

Even though both concrete compositions were similar, a big difference in the compressive strength was detected. This effect might be attributed to higher air content in the GF mix compared with the SF series. Moreover, taking into account the diameter, specific weight and the different contents of fibres, the amount of glass fibres was about 100 times higher than steel fibres. This led to more interfacial transition zones which affected the compressive strength.

Table 1 Mix proportioning of concrete

Material	SF (kg/m^3)	GF (kg/m^3)
Cement 52, 5R	800	800
Silica sand 0/2	1161	1161
Filler ($CaCO_3$)	200	200
Water	228	236
Nanosilica	40	40
Superplasticizer	30	30
Steel fibres	150	–
Glass fibres	–	44

Table 2 Properties of the fibres

Characteristics	Steel fibres	Glass fibres
Geometry	Straight	Filament
Length (mm)	13	13
Diameter (mm)	0.2	0.018
Tensile strength (MPa)	2300–2500	1400
Elastic modulus (GPa)	210	74
Specific weight (g/cm^3)	7.75–8.05	2.7

Table 3 Average mechanical properties of the concrete

Parameters (MPa)	SF		GF	
	Average	CV (%)	Average	CV (%)
E_{cm}	34970	3.16	31137	2.72
f_{cm}	102.79	1.54	73.59	3.14
f_{LOP}	7.24	16.59	7.68	11.02
f_{R1}	10.41	9.36	5.03	15.20
f_{R2}	9.79	8.63	1.77	14.41
f_{R3}	8.13	8.88	0.82	13.96
f_{R4}	6.95	11.32	0.46	18.47

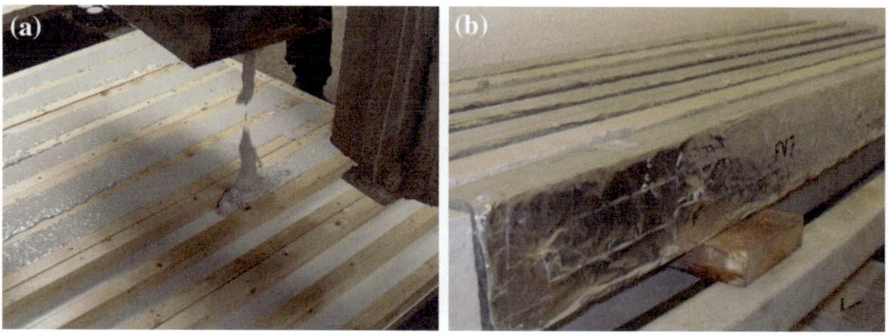

Fig. 3 **a** Pouring of the beams and **b** wrapped beam in moist room

To analyse the effect of creep, 12 beams of 40 × 80 × 1200 mm per each dosage were manufactured. The concrete was produced in a vertical axis mixer and directly poured from the skip to the moulds (Fig. 3a). To prevent the loss of superficial water and a consequent early shrinkage a curing layer was sprayed over the free surface of the concrete. All the specimens were demoulded 24 h after the fabrication and were immediately stored in a moist room with a controlled temperature of 20 °C and a relative humidity of 100 % (Fig. 3b). Half of the beams corresponding to each series were wrapped in aluminium tape to analyse the influence of restricting the humidity transfer during the curing.

3.2 *Pre-cracking and Loading Sequence*

The beams were pre-cracked applying load at the mid-span according to the three-point test setup as depicted in Fig. 4. Notches were not performed to avoid reduction of the section. The distance between supports was 500 mm and the load was transmitted to the beam at a constant rate of 0.3 mm/min. When the beam reached the flexural strength, the loading operation was immediately stopped. This

Fig. 4 Pre-cracking of the beam in a three-point configuration

point was identified by means of both the load-displacement slope variation and the load dropping occurred at that instant. Even though the common procedure in creep tests is arriving to a certain crack width or deflection [6], the procedure here described was systematically followed for all specimens. Moreover, it is worth to mention that no additional damage was undergone by any sample, since in no case the deflection exceeded 1 mm.

Since the span changed from 500 mm at the load press to 1100 mm at the frame, the loads P in each case needed to be recalculated as described in Eq. (1) by keeping the bending moment constant for both lengths. As a result, the load to produce the same moment in the beam of 1100 mm decreased with respect to the load for a beam of 500 mm.

$$P_{1100\mathrm{mm}} = \frac{l_{500\mathrm{mm}}}{l_{1100\mathrm{mm}}} \cdot P_{500\mathrm{mm}} \quad (1)$$

Nevertheless, a different behaviour between wrapped and non-wrapped specimens after the crack strength was detected. In non-wrapped specimens there was a small drop of the strength after the crack load, followed by an increase of the strength even over the peak. In wrapped specimens a sudden drop of the load occurred immediately after the peak load was reached, showing no significant increase of the load from this point onwards. Moreover, higher peak values were obtained in wrapped beams, possibly due to the additional effect of the aluminium since it was not removed.

As a result, two different strategies were defined for loading the beams during the creep test depending on whether the beams were wrapped or non-wrapped. As depicted in the schemes of Table 4, the reference load for non-wrapped specimens was the cracking load (P_{cr}), whereas for wrapped specimens the reference load was that in which the major cracks stabilize and the residual strength appears (P_{cr}^{*}). The load for each beam in the creep test was calculated on the basis of these data.

For non-wrapped specimens the creep load was defined as 50 % of P_{cr} from the beginning of the test, and no changes on the load were later performed. Wrapped specimens were initially loaded at 25 % of P_{cr}^{*} and after a period of 15 days, the

Table 4 Different load levels (kN) in non-wrapped and wrapped HPFRC beams

	Scheme	Beam	L = 500 mm		L = 1100 mm		
			P_{cr}	P^*_{cr}	t_0 (0 days)	t_1 (15 days)	t_2 (30 days)
Non-wrapped	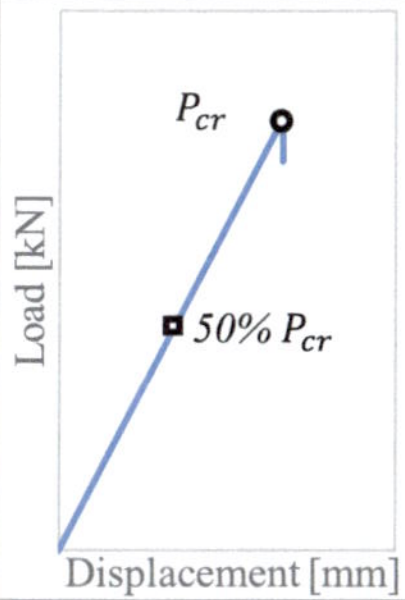	SF1	1.43	–	0.33 (50 $\%P_{cr}$)	–	–
		SF2	1.93	–	0.44 (50 $\%P_{cr}$)	–	–
		SF3	1.61	–	0.37 (50 $\%P_{cr}$)	–	–
		GF1	1.33	–	0.30 (50 $\%P_{cr}$)	–	–
		GF2	1.33	–	0.30 (50 $\%P_{cr}$)	–	–
		GF3	1.52	–	0.34 (50 $\%P_{cr}$)	–	–
Wrapped	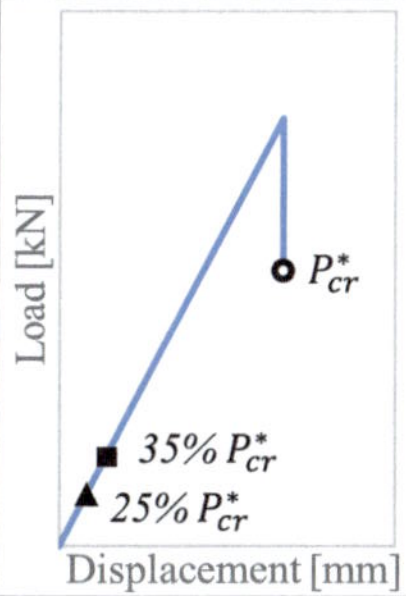	SF6	2.29	1.25	0.14 (25 $\%P^*_{cr}$)	0.20 (35 $\%P^*_{cr}$)	0.52 (50 $\%P_{cr}$)
		SF7	2.57	1.49	0.17 (25 $\%P^*_{cr}$)	0.23 (35 $\%P^*_{cr}$)	0.58 (50 $\%P_{cr}$)
		SF8	2.49	1.58	0.18 (25 $\%P^*_{cr}$)	0.24 (35 $\%P^*_{cr}$)	0.57 (50 $\%P_{cr}$)
		GF6	1.86	0.80	0.42 (50 $\%P_{cr}$)	–	–
		GF7	2.19	1.46	0.17 (25 $\%P^*_{cr}$)	0.23 (35 $\%P^*_{cr}$)	–
		GF8	1.79	1.00	0.12 (25 $\%P^*_{cr}$)	0.16 (35 $\%P^*_{cr}$)	–

load was changed to 35 % of P^*_{cr}. An exception was made with GF6, which was initially loaded with 50 % of P_{cr} but immediately collapsed. At 30 days, to compare the creep between non-wrapped and wrapped beams, the creep load of steel fibre reinforced wrapped beams was pushed to the limit and it was increased to 50 % of P_{cr}.

4 Results

Creep was evaluated measuring the vertical displacement at the mid-span of the beams. In this section, the evolution of the deflection and the evolution of the creep coefficient are analysed. The differences on these two parameters are discussed according to the influence of the type of fibre and curing method.

4.1 Evolution of the Deflection

The total deflection at a time t, $\delta_{tot}(t)$, is the direct sum of the initial deflection, $\delta(t_o)$, due to the instant effect of the loading of the beams and the deflection due to creep $\delta_\varphi(t)$. In Fig. 5 it is shown the deflection due to creep of each beam with steel fibres (SF) or glass fibres (GF), as well as their average value. The results also reveal the differences between non-wrapped and wrapped specimens. The data of non-wrapped beams are shown for a time of 150 days as no changes on the load level were introduced within that period. In wrapped specimens the level loads were changed two times. Only data until the day 30 appear, since during that period the two level loads were the same for both types of fibre.

Figure 5a indicates that both SF and GF present a similar $\delta_\varphi(t)$ until the day 30. At this point, when the deflection in GF is only 2.9 % higher than in SF, the trend starts to change and the deflection detected in the GF beams gradually increases

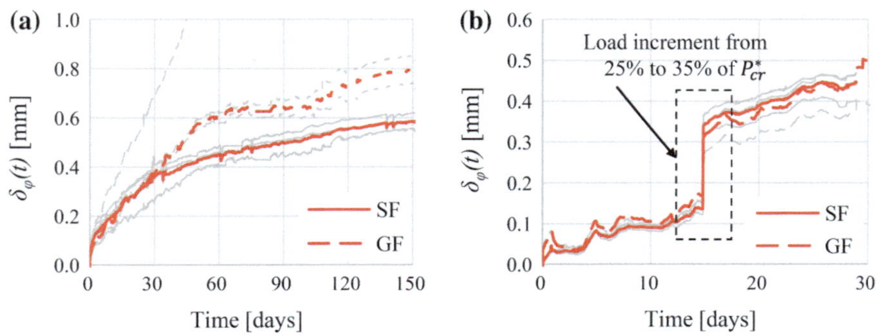

Fig. 5 Average creep deflection in **a** non-wrapped and **b** wrapped beams

Table 5 Deflection due to creep at 15 and 30 days

Deflection (mm)	Non-wrapped (50 % P_{cr})		Wrapped (35 % P^*_{cr})	
	SF	GF	SF	GF
$\delta_\varphi(15)$	0.254	0.241	0.338	0.306
$\delta_\varphi(30)$	0.363	0.374	0.447	0.452

with respect to SF. At 150 days, the deflection due to creep in GF is 36 % higher than in SF. These results highlight that the biggest differences between the two types of fibres in terms of $\delta_\varphi(t)$ are evident from 30 days onwards.

One of the beams reinforced with glass fibres (GF2) revealed an unexpected behaviour. Since the beginning of the test the deflection was significantly higher than the deflection of the other GF beams. This inconsistency with respect to the other results may be due to a defective manufacture of this individual beam, since its pre-cracking and loading procedure was identical to the rest of the beams.

The performance of wrapped specimens is gathered in Fig. 5b. As it happened in non-wrapped specimens, the average deflection of the beams resulted to be very similar regardless of the type of fibre used as reinforcement. Moreover, when the load was increased at the day 15 from 25 to 35 % of P^*_{cr} the deflection in SF increased from 0.136 to 0.338 mm, whereas in GF this grew from 0.169 to 0.306 mm. At 30 days, the average difference of $\delta_\varphi(t)$ between beams with SF and GF was only 1 %, thus excluding the type of fibre as a factor influencing the deflection in such low load levels. Further study should be undertaken to identify at which load level the fibres become the distinguishing feature affecting $\delta_\varphi(t)$.

Comparing the results gathered in Table 5 regarding the curing method between beams, it was noticed that $\delta_\varphi(t)$ at 15 and 30 days was higher in wrapped specimens when the load level was 35 % of P^*_{cr}. At 15 days, the deflection due to creep in SF and GF was 25 and 21 % higher than in non-wrapped specimens, respectively. At 30 days, these percentages decreased respectively to 19 and 17 %, still remaining the deflection in wrapped specimens greater than in non-wrapped.

The latter results seem to be contradictory in relation to the load level since the load in wrapped specimens was in every case lower than in non-wrapped. This could be attributed to the different curing process produced by the lack of external water contribution during the set of concrete. This might have caused a lower amount of hydrated cement paste and, consequently, a lower elastic modulus. In this case, deformation would be greater in wrapped specimens in comparison to others with a higher degree of hydration and higher elastic modulus.

4.2 Evolution of the Creep Coefficient

The creep coefficient at a defined time t, $\varphi_c(t)$, may be calculated as the ratio between the deflection produced by creep $\delta_\varphi(t)$ and $\delta(t_o)$, as indicated in Eq. (2).

Although this is not the classical formulation of creep, previous studies report the possibility of determining the creep coefficient in this way when the deformation is not directly measured.

$$\varphi_c(t) = \frac{\delta_\varphi(t)}{\delta(t_0)} \quad (2)$$

Table 6 summarizes the creep coefficients at 15, 30, 90 and 150 days as well as the increments between these periods of time and Fig. 6 shows the whole evolution of the creep coefficient through time. The results of the beams presented in Table 6 correspond to the specimens without aluminium wrapping and did not suffer any load variations during the creep test. No results of GF2 are available from the day 47 onwards as a result of its failure.

The values of the creep coefficient were found to be in all cases higher in glass fibre reinforced beams than in steel fibre beams. At the day 15, GF3 presented the lowest creep coefficient of glass fibre beams, which was the same as the highest creep coefficient of steel fibres obtained in SF2. As reported in previous investigations [12], this effect may be attributed to the different elastic modulus of the two different concretes, since the higher capacity of deformation of glass fibres is a

Table 6 Creep coefficients in non-wrapped beams

Beam	$\varphi_c(15)$	$\varphi_c(30)$	$\varphi_c(90)$	$\varphi_c(150)$	$\varphi_c(15–30)$ (%)	$\varphi_c(30–90)$	$\varphi_c(90–150)$
SF1	0.25	0.35	0.61	0.75	44.5	71.3 %	23.6 %
SF2	0.26	0.36	0.47	0.54	39.2	31.5 %	15.0 %
SF3	0.21	0.30	0.37	0.42	41.5	23.9 %	15.1 %
GF1	0.71	1.09	1.88	2.27	52.9	72.4 %	20.8 %
GF2	0.45	0.73	–	–	61.8	–	–
GF3	0.26	0.40	0.70	0.88	50.4	75.4 %	26.0 %

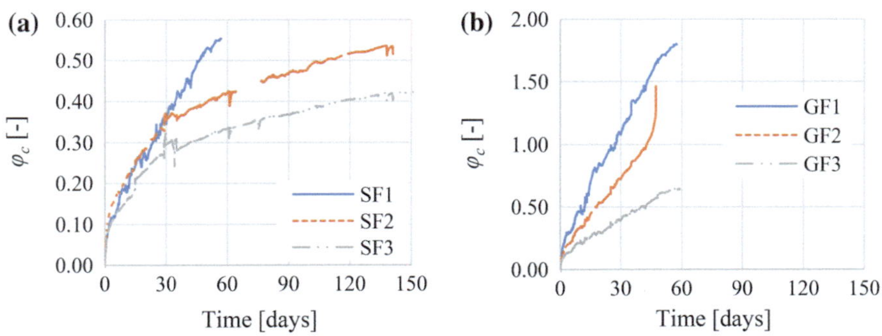

Fig. 6 Relationship between time and creep coefficient in non-wrapped beams

result of the lower elastic modulus and tensile strength when compared to steel fibres (see Table 2).

In the case of SF, the greatest increments of the creep coefficient were produced between the days 15 and 30, whereas in GF the biggest increment occurred between the days 30 and 90. Nevertheless, in both types of beam, from the day 90 the creep coefficient reduces its value and the increments between 90 and 150 days experience a drop in comparison with those obtained for 30–90 days. Longer periods for the creep test would provide further information to make long term predictions and consider the possibility of sudden failure of fibres.

The variability in the results between beams is also noticeable. The variation of the creep coefficient in SF for days 15 and 30 was around 10 %, whereas the CV in GF at the same days was approximately 47 %. These CV increased at days 90 and 150, achieving an average value of 27 % in SF and 63 % in GF.

The creep coefficients of the wrapped beams are shown in Table 7 and their evolution during the test including the load changes in Fig. 7. Regarding the values gathered in Table 7, given that these beams experienced different load levels, the creep coefficients are shown with respect to the load level at which they were obtained. At days 15 and 30 two coefficients are presented since those were the days when the load levels were changed. Two additional coefficients (days 90 and 150) are also presented, these only for GF yet beams with SF collapsed between 2 and 4 days after the load was increased to 50 % of P_{cr} at the day 30.

The creep coefficient at 15 days for the same load level was in average two times roughly higher in GF than in SF. At 15 and 30 days with a load level of 35 % P^*_{cr}, φ_c of GF was found to be again higher than the coefficient for SF. The coefficient kept growing but at 150 days it decreased around 16 % in comparison with the value obtained at 90 days.

However, the biggest increments were detected at the load level changes rather than those produced as a result of the creep. An example of this situation is GF6, which accidentally failed as a result of an excessive load level. Regarding the rest of the beams, an increase of the load from 25 % of P^*_{cr} to 35 % produced a boost of around 150 % in SF, whereas for GF that percentage was approximately 80 %. When in SF the load was changed at the day 30 to 50 % of P_{cr} the increments were of 315 % in SF6-SF7 and 206 % in SF8, producing the collapse of the beams between 2 and 4 days later.

Table 7 Creep coefficients in wrapped beams

Beam	$\varphi_c(15)$ (25 % P^*_{cr})	$\varphi_c(15)$ (35 % P^*_{cr})	$\varphi_c(30)$ (35 % P^*_{cr})	$\varphi_c(30)$ (50 % P_{cr})	$\varphi_c(90)$ (35 % P^*_{cr})	$\varphi_c(150)$ (35 % P^*_{cr})
SF6	0.19	0.50	0.63	2.63	–	–
SF7	0.24	0.65	0.85	3.51	–	–
SF8	0.35	0.77	1.08	3.30	–	–
GF7	0.58	1.10	1.67	–	2.12	1.73
GF8	0.52	0.89	1.27	–	1.38	1.17

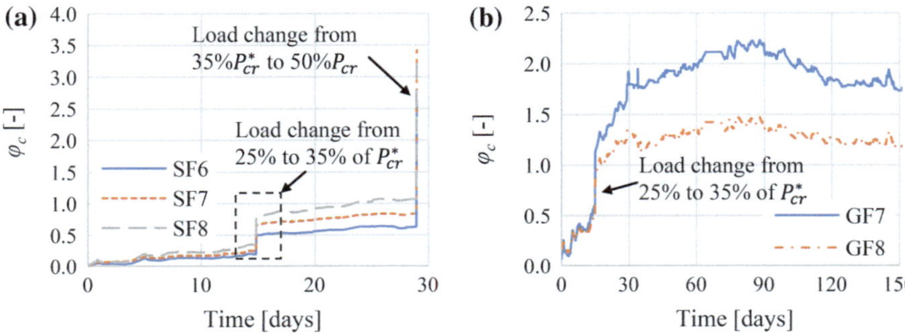

Fig. 7 Evolution of creep coefficient in wrapped beams

5 Conclusions

The main goal of the current study was to design and evaluate a new type of frame to analyse creep in HPFRC. Additionally, 12 HPFRC beams were tested measuring the deflection at the mid-span in a pre-cracked state. Based on the performance of the frame designed and constructed and the results obtained during the creep test, several conclusions may be drawn:

- A new frame to analyse the flexural creep of FRC elements has been conceived, this allowing the possibility of testing at the same time several elements.
- It was confirmed that the load of each beam may be individually customized without interfering on the other specimens. This allows studying several variables such as the type of fibre or the load level of each individual specimen.
- The experimental campaign conducted verified the applicability of the frame on testing creep in HPFRC beams. The loads were able to be kept constant during all the test period of time and they could also be changed when necessary to adapt new load levels.
- The load level should be established depending on the behaviour after the flexural strength to avoid sudden collapse. Beams curing without external contribution of water presented lower residual strength and, therefore, the load level had to be reduced.
- At low load levels (25–35 % P^*_{cr}) the type of fibre had no significant effect on the deflection. Not even the increase of the load level from 25 to 35 % P^*_{cr} produced notable differences between the types of fibre reinforcement.
- The deflection in wrapped specimens was higher than in non-wrapped. This was reflected at the case when wrapped specimens were loaded with 35 % P^*_{cr} and the load in non-wrapped was 50 % of P_{cr}.
- The creep coefficient in GF was in all cases higher than in SF. This might be a consequence of the lower residual strength (f_{Ri}) in GF.

Acknowledgments The authors acknowledge the collaboration of ESCOFET S.A. in the performance of the experimental program. The first author also acknowledges the grant FPU provided by the Spanish Ministry of Education, Culture and Sports.

References

1. Walraven, J.: High performance fiber reinforced concrete: progress in knowledge and design codes. Mater. Struct. **42**(9), 1260–1274 (2009)
2. Bernard, E. S.: Creep of cracked fibre reinforced shotcrete panels. In: Shotcrete: More Engineering Developments, Proceedings of the Second International Conference on Engineering in Shotcrete, Cairns, Australia, pp. 47–57. Taylor and Francis Group, London (2004)
3. Barragán, B. E., Zerbino, R. L.: Creep behaviour of cracked steel fibre reinforced concrete beams. In: Proceedings of the 7th International RILEM Symposium on Fibre Reinforced Concrete: Design and Applications (BEFIB 2008), Chennai, India, pp. 577–586 (2008)
4. Serna Ros, P., Bossio, M. E., Zerbino, R., Martí Vargas, J. R.: Fluencia y propiedades residuales de hormigones autocompactantes con fibras expuestos en estado fisurado a diferentes condiciones ambientales, 3° Congreso Iberoamericano sobre hormigón autocompactante, pp. 221–230 (2012)
5. Abrishambaf, A., Barros, J.A.O., Cunha, V.M.C.F.: Time-dependent flexural behaviour of cracked steel fibre reinforced self-compacting concrete panels. Cem. Conc. Res. **72**, 21–36 (2015)
6. Arango, S.E., Serna, P., Martí-Vargas, J.R., García-Taengua, E.: A test method to characterize flexural creep behaviour of pre-cracked FRC specimens. Exp. Mech. **52**(8), 1067–1078 (2012)
7. Pujadas, P.: Caracterización y diseño del hormigón reforzado con fibras plásticas. PhD. Thesis, Universitat Politècnica de Catalunya (2013)
8. Blanco, A.: Characterization and modelling of SFRC elements. PhD. Thesis, Universitat Politècnica de Catalunya (2013)
9. CEN, EN 12390-13:2014: Testing hardened concrete—Part 13: Determination of secant modulus of elasticity in compression. European Committee for Standardization, Brussels (2014)
10. CEN, EN 12390-3:2009: Testing hardened concrete—Part 3: Compressive strength of test specimens. European Committee for Standardization, Brussels (2009)
11. CEN, EN 14651:2007: Test method for metallic fibre concrete. Measuring the flexural tensile strength (limit of proportionality (LOP), residual). European Committee for Standardization, Brussels (2007)
12. Zhao, O., Yu, J., Geng, G., Jiang, J., Liu, X.: Effect of fiber types on creep behavior of concrete. Constr. Build. Mater. **105**, 416–422 (2016)

Time-Dependent Flexural Behaviour of SFRSCC Elements

Vítor M.C.F. Cunha, Joaquim A.O. Barros and Amin Abrishambaf

Abstract This work presents and discusses the long-term behaviour of pre-cracked steel fibre reinforced self-compacting concrete (SFRSCC) laminar structures of relatively small thickness. One hundred and twelve prismatic specimens were extracted from a SFRSCC panel. These specimens were notched with different orientations regarding to the expected SFRSCC flow direction, and were tested under four-point flexural sustained loading conditions. The influence of the following parameters on the creep behaviour was studied: initial crack opening level (0.3 and 0.5 mm), applied stress level, fibre orientation/dispersion, and distance from the casting point. Moreover, to evaluate the effect of the long-term residual crack opening on the flexural post-cracking strength, as well as on the secondary stiffness, a series of instantaneous monotonic and cyclic tests were carried out, and the corresponding force vs crack tip opening displacement (*F*–CTOD) curves were compared to the ones obtained by assembling the *F*–CTOD curves determined in the pre-crack monotonic tests, creep tests and post-creep monotonic tests. Finally, based on the results obtained from the creep tests, an equation was proposed to predict the creep coefficient for the developed SFRSCC.

Keywords Creep · Steel fibres · Self-compacting concrete · Fibre orientation · Flexural residual strength

1 Introduction

The instantaneous post-cracking behaviour of fibre reinforced concrete, FRC, has been widely studied in the last decades, either by direct or indirect methods, by e.g. [1–5]. Nevertheless, the knowledge available in literature regarding the long-term response of FRC is still somehow scarce. Moreover, the major part of the studies

V.M.C.F. Cunha (✉) · J.A.O. Barros · A. Abrishambaf
Department of Civil Engineering, School of Engineering, ISISE, University of Minho, Guimarães, Portugal
e-mail: vcunha@civil.uminho.pt

P. Serna et al. (eds.), *Creep Behaviour in Cracked Sections of Fibre Reinforced Concrete*, RILEM Bookseries 14,
DOI 10.1007/978-94-024-1001-3_11

mainly address the creep behaviour of concrete reinforced with synthetic fibres, by e.g. [6–8], which is understandable due to the higher susceptibility of these fibres to thermo-hygrometric environmental conditions, but the increasing use of steel FRC (SFRC) for structural applications brings renovated interested in this topic. Thus, to cope with these debilities, recently, some works regarding the creep behaviour of SFRC have been presented, by e.g. [9–12]. It is worth noting that the aforementioned studies were focused mainly on prismatic specimens, which in the case of using self-compacting concrete mixtures have a distinct fibre distribution and orientation from the one of planar structures [13–15]. Despite the availability of some standards for designing SFRC structures [16, 17], they still do not envisage the long-term behaviour of SFRC in cracked conditions. Therefore, information regarding the long-term behaviour of cracked SFRC elements, particularly of planar structures (like the outer layers of sandwich panels, thin slabs and shells), is still limited. Consequently, understanding the behaviour of cracked SFRC elements under a sustained load will help towards a more rational design and accurate prediction of the composite behaviour under serviceability conditions. To sum it up, it is important to evaluate the concrete capability to maintain the crack opening width relatively low under a sustained load, in order to guarantee the effectiveness of fibre reinforcement under serviceability conditions.

The long-term behaviour of pre-cracked steel fibre reinforced self-compacting concrete (SFRSCC) laminar structures (of relatively small thickness) will be presented and discussed. One hundred and twelve prismatic specimens were extracted from a SFRSCC panel. These specimens were notched with different orientations regarding to the expected SFRSCC flow direction, and were tested under four-point flexural sustained loading conditions. The influence of the following parameters on the creep behaviour was studied: initial crack opening level, w_{cr} (0.3 and 0.5 mm); applied load level, (50–100 %), fibre orientation/dispersion, and distance from the casting point. Moreover, to evaluate the effect of the long-term residual crack opening on the flexural post-cracking strength, as well as on the secondary stiffness, a series of instantaneous monotonic and cyclic tests were carried out, and the corresponding force vs crack tip opening displacement (*F*–CTOD) curves were compared to the ones obtained by assembling the *F*–CTOD curves determined in the pre-crack monotonic tests, creep tests and post-creep monotonic tests. Finally, based on the results obtained from the creep tests, an equation was proposed to predict the creep coefficient for the developed SFRSCC.

2 Experimental Programme

2.1 Concrete Composition

Table 1 comprises the composition of the developed SFRSCC. W/C abbreviates water (W) to cement (C) ratio. Superplasticizer Sika® 3005 (SP) was used to assure

Table 1 Steel fibre reinforced self-compacting concrete composition per m^3

C (kg)	W (kg)	W/C (–)	SP (kg)	Filler (kg)	FS (kg)	CS (kg)	CA (kg)	Cf (kg)
413	124	0.30	7.83	353	237	710	590	60

self-compactibility requirements. A crushed granite coarse aggregate (CA) with a maximum aggregate size of 12 mm, a fine sand (FS) and coarse sand (CS) were adopted. A content (Cf) of 60 kg/m^3 of hooked steel fibres with a length, *lf*, of 33 mm, a diameter of 0.55 mm, *df*; and a yield stress of 1100 MPa was used. The fresh state behaviour of concrete was determined by the Abrams cone slump test in the inverted position according to [18], having been obtained a spread diameter of approximately 590 mm.

The compressive strength was assessed by testing 6 cylinders with a diameter of 150 mm and a height of 300 mm at the age of 28 days. After casting, the specimens were stored in a climatic chamber room at constant temperature of 20 °C and relative humidity of 60 %. An average compressive strength and an average young modulus of 72 MPa and 42 GPa was obtained, respectively, with a coefficient of variation (CoV) of 8.2 and 0.3 %, respectively.

2.2 *Specimens*

The SFRSCC specimens used for assessing the creep behaviour in cracked conditions were extracted from a panel with 1.5 × 1.5 m^2 in plan and 0.06 m of thickness. The panel was cast from its geometric centre. A total of one hundred and twelve specimens with the dimensions of 0.24 × 0.06 × 0.06 m^3 were extracted from distinct locations. The specimen's notch depth and thickness were 10 and 2 mm, respectively. The orientation of the extracted prismatic specimens within the panel was established having in mind the expected concrete flow direction, see Fig. 1, considering β as the angle between the direction of the concrete flow and the notched plane direction. Four intervals with different β were defined, i.e. [0–15°[, [15–45°[, [45–75°[and [75–90°]. In Fig. 1a, the light grey solid hatched specimens were used for assessing the long-term behaviour, whereas the rest were tested under instantaneous monotonic load conditions at the same age of the specimens cracked for the creep test program.

2.3 *Test Setup*

Figure 2a depicts the adopted test setup for obtaining the instantaneous monotonic behaviour, the specimens' pre-cracking and post-creep behaviour. A universal testing rig with 50 kN load capacity was used. The tests were performed in

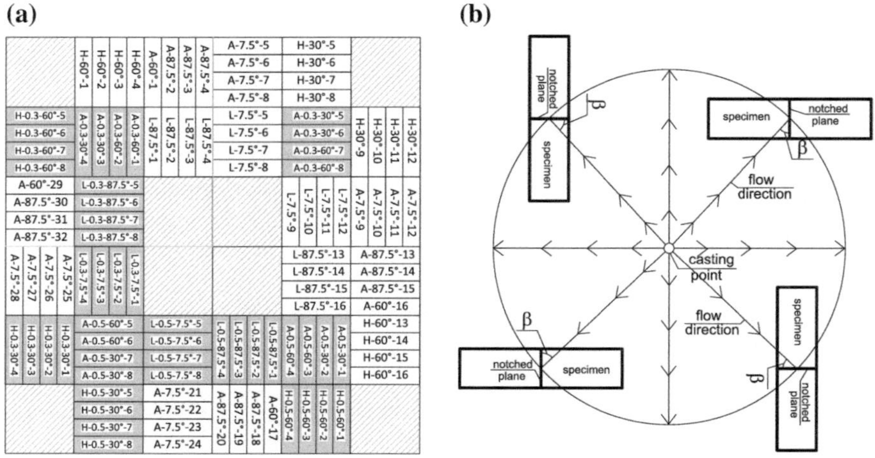

Fig. 1 **a** Specimen's extraction location (dimensions in plan of 1.5 × 1.5 m^2); **b** definition of the β angle

closed-loop control with a displacement rate of 0.002 mm/s. These tests were carried out following the recommendation of [16]. The bottom face of the specimen in the test setup corresponds to a lateral surface of the extracted beam from the panel, i.e. the beam was rotated 90° along its longitudinal axis for testing. To assess an eventual asymmetric crack opening, due to fibre segregation along the panel's thickness, CMOD was measured by installing three LVDTs on the beam's bottom surface, see Fig. 2a.

On the other hand, Fig. 2b shows the adopted test setup for assessing the creep crack opening width–time relationship (Fig. 3a). The test was executed under force control. The crack opening width was measured with one LVDT installed at the middle of the notched surface. The creep four-point bending tests were performed in pre-cracked specimens, namely, 0.3 mm (recommended for serviceability limit state) and 0.5 mm (crack opening width corresponding to the one used to compute the residual flexural strength, $f_{R,1}$ [17]). After finalizing the pre-cracking procedure, the specimens were carefully moved to the climate chamber room, at a temperature of 20 ± 0.5 °C and a 60 ± 5 % relative humidity. After the initiation of the creep test, the value of the applied load, F_a, was fixed and maintained constant until the crack opening width was stabilized (i.e. crack opening increment value smaller than 1 μm for three consecutive days). At the end of the creep tests, the specimens were unloaded, but the data acquisition system was kept active for at least one week, enabling to record the crack width closure due to the creep recovery process. Afterwards, the post-creep behaviour was assessed. Finally, the assembled *F-w* curves were obtained with the individual *F-w* curves from the pre-cracking, creep and post-creep responses, as shown in Fig. 3b. Further details regarding the adopted test setups and procedure to obtain the assembled curve can be found elsewhere [12].

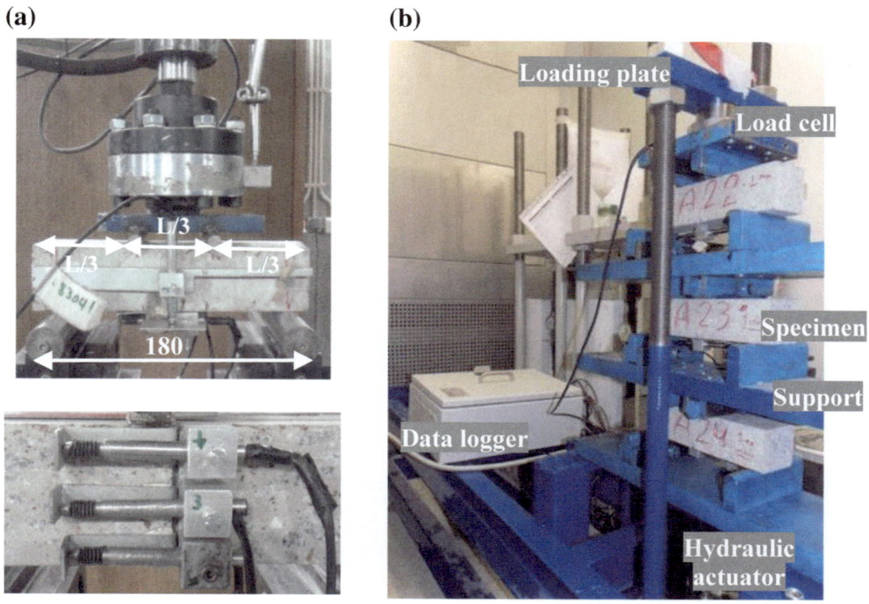

Fig. 2 Test setup of the four-point bending tests: **a** Monotonic (pre-cracking and post-creep); **b** long-term creep

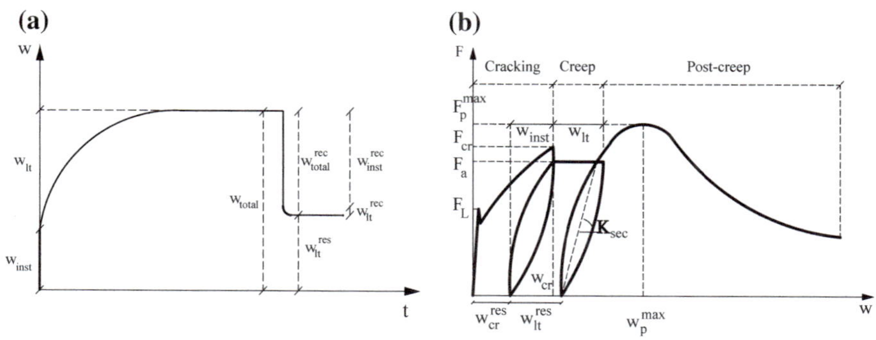

Fig. 3 **a** Crack opening-time relationship; **b** assembled long-term force-crack width curve

2.4 *Assessment of the Fibre Orientation and Distribution*

The distribution and orientation of fibres in the SFRSCC panel was assessed by an image analysis procedure [15, 19]. The image analysis was carried out on a plane parallel to the notch plane with an offset equal to half the length of the used fibre. The grinded plane was obtained by cutting the specimens after the monotonic four-point bending tests have been carried out. For each series (β = [0–15°[, [15–45°[, [45–75°[and [75–90°]) five specimens from distinct panel locations were selected. After computation and analysis of the image technique results, the following parameters were derived out [15]: (i) the number of fibres per unit area, N^f; (ii) fibre orientation factor, η_θ; (iii) Fibre segregation parameter, ξ_{seg}. The number of the effective fibres per unit area, N^f_{eff}, was also determined by manually counting the fibres that intersected the notched plane of the tested beams and also had the hooked end deformed (in order to be considered a fibre that provided effective reinforcement).

3 Results and Discussion

3.1 *Fibre Orientation and Distribution*

Table 2 includes the fibre distribution parameters obtained by image analysis. N^f and N^f_{eff} were higher in the specimens with the notched plane parallel to the expected concrete flow direction. These values decreased as the notched plane rotated towards to the perpendicular position regarding the flow direction (Fig. 1b). For the case of the notched plane orientation β = [0–15°[, N^f_{eff} was approximately 76, 156 and 686 % higher than when β was between [15–45°[, [45–75°[and [75–90°] intervals, respectively. A higher value for the fibre orientation factor, η_θ, was also obtained for the series with a β = [0–15°[, having been 8, 20 and 30 % higher than the one obtained for the series [15–45°[, [45–75°[and [75–90°], respectively. This could be ascribed to a preferential fibre alignment, which was influenced by the flowability of concrete that induces the fibres to be reoriented and remain preferentially perpendicular to the concrete flow direction. The fibre segregation parameter, ξ_{seg}, was slightly higher than 0.5 indicating the occurrence of a slight segregation of the fibres, caused by the highest specific weight of the steel fibres amongst the constituents of the SFRSCC.

3.2 *Long-Term Four-Point Bending Test*

Figure 4 shows both the single and average crack mouth opening displacement *vs* time relationships, CMOD—*t*, for the two pre-crack widths investigated

Table 2 Fibre distribution parameters obtained after the monotonic four-point bending tests

Series	N^f (fibres/cm^2)		N^f_{eff} (fibres/cm^2)		η_θ (–)		ξ_{seg} (–)	
	Avg.	CoV (%)	Avg.	CoV (%)	Avg.	CoV (%)	Avg.	CoV (%)
β = [0–15°]	2.12	21.1	1.18	27.5	0.875	1.0	0.52	2.5
β = [15–45°]	1.76	11.8	0.67	28.6	0.807	4.4	0.52	2.6
β = [45–75°]	1.53	25.9	0.46	19.6	0.730	6.4	0.58	8.8
β = [75–90°]	0.87	21.8	0.15	78.4	0.672	12.9	0.62	15.2

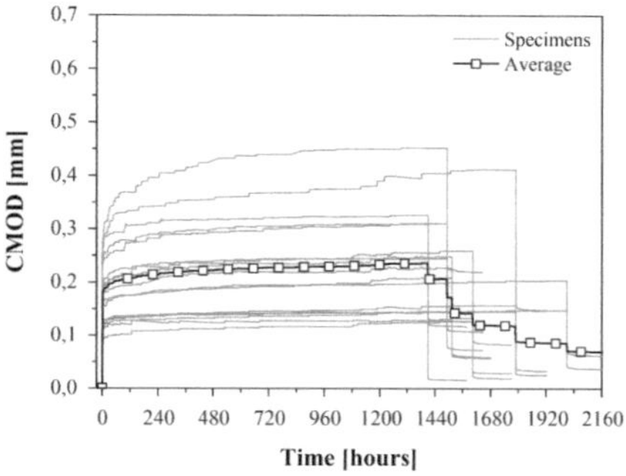

(a) $w_{cr} = 0.3$ mm

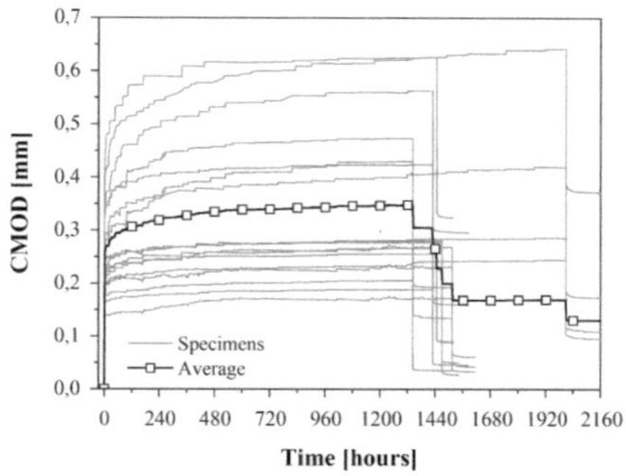

(b) $w_{cr} = 0.5$ mm

Fig. 4 Crack mouth opening displacement, CMOD, versus time relationships

(i.e. w_{cr} = 0.3 and 0.5 mm) obtained during the long-term four-point bending tests. During the test execution, none of the specimens entered into the tertiary creep stage, in which occurs failure due to creep. In general, the evolution of the w_{cr} = 0.3 and 0.5 mm responses with time was similar, and the higher CMOD levels registered for the w_{cr} = 0.5 mm series was a direct consequence of the higher damage at the fibre/matrix interface imposed by the higher pre-crack width. Note that the high scatter in the specimens' responses is ascribed to distinct applied load levels, notch orientations and locations.

3.3 *Influence of the Loading Ratio on the Creep Behaviour*

To analyse the long-term behaviour of the cracked SFRSCC, the creep coefficient parameter in the creep stage (φ^C) was calculated by [20]:

$$\varphi^C = w_{lt}/w_{inst} \tag{1}$$

where w_{lt} and w_{inst} are the long-term crack width and instantaneous crack opening width at the time of loading, respectively, see Fig. 3a.

Figure 5 depicts the variation of the creep coefficient for different loading ratios of F_a/F_{cr}, see Fig. 3b. This coefficient has increased with the load level ratio. Considering the influence of pre-crack width, the creep coefficient in w_{cr} = 0.5 mm series has increased with the F_a/F_{cr} at a higher rate than in the case of the series of w_{cr} = 0.3 mm. By considering the micromechanical behaviour of a single fibre, the w_{cr} = 0.5 series was submitted to higher damage level of fibre/matrix interface, which led to a higher increase of the crack opening width under a creep load [12]. This influence was more significant for a load level of 100 %. Therefore, sustained loads near to the maximum instantaneous bearing capacity of the specimen can lead to a significant increase of the creep coefficient values.

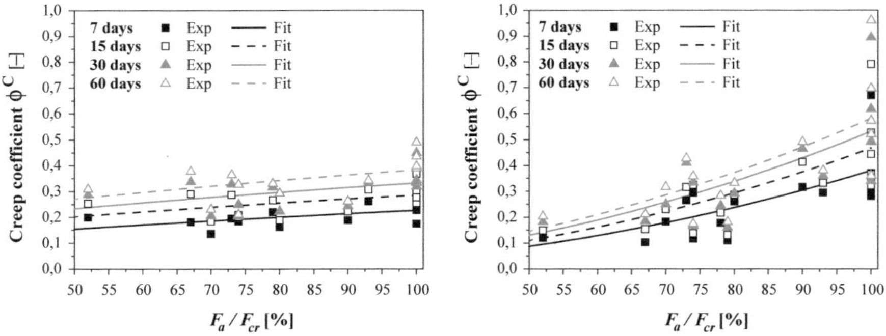

(a) $w_{cr} = 0.3$ mm (b) $w_{cr} = 0.5$ mm

Fig. 5 Relationship between the creep coefficient and the loading ratio, F_a/F_{cr}

3.4 Influence of Notch Plane Orientation on the Creep Behaviour

Figure 6 presents the influence of the notched plane orientation (regarding the flow direction of SFRSCC) on the relationship of creep coefficient versus time. These relationships were determined by averaging the response of the specimens with $F_a/F_{cr} = 100$ %. For the $w_{cr} = 0.3$ mm series, the variation of creep with time was slightly affected by the direction of the notched plane regarding the expected concrete flow. On the other hand for $w_{cr} = 0.5$ mm series, this influence was more significant. In the case of the β = [0–15°[series, the creep coefficient at the end of two months was 13 and 31 % higher than in β = [75–90°] for the $w_{cr} = 0.3$ and 0.5 mm series, respectively. It is worth noting that the β = [0–15°[specimens contain more effective fibres perpendicular to the crack plane, which were

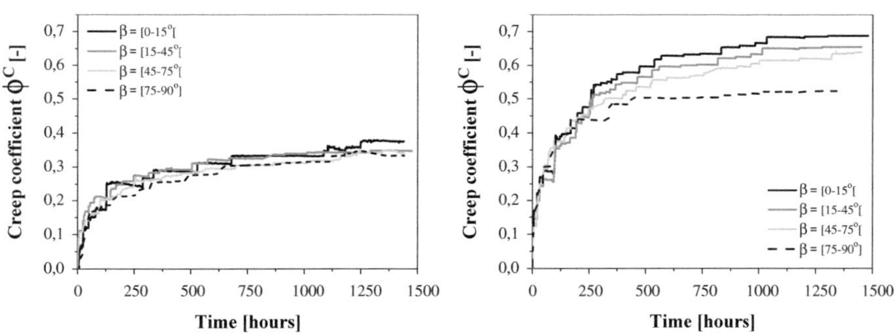

(a) $w_{cr} = 0.3$ mm (b) $w_{cr} = 0.5$ mm

Fig. 6 Creep coefficient versus time for different orientations of the notched plane

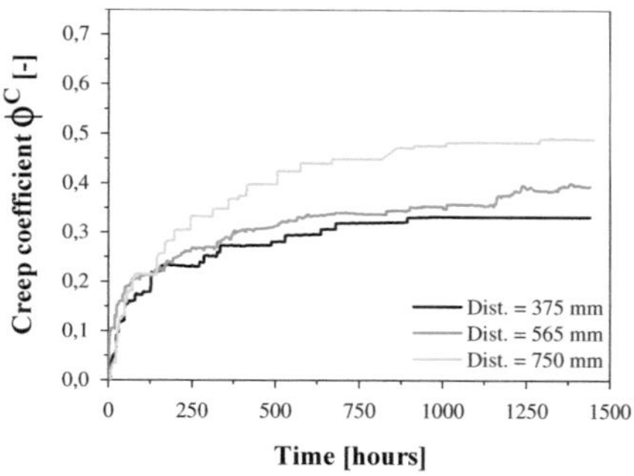

(a) $w_{cr} = 0.3$ mm

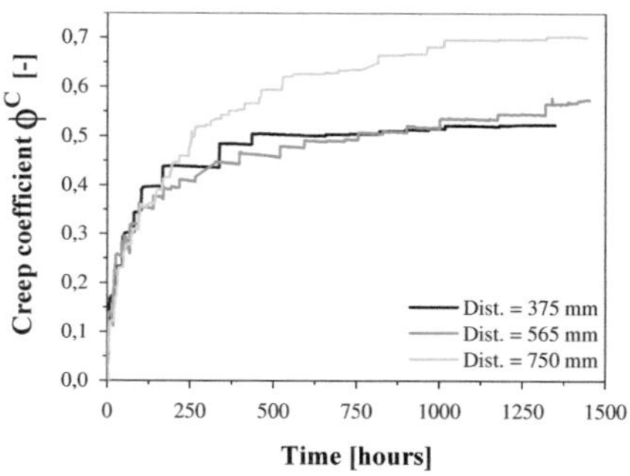

(b) $w_{cr} = 0.5$ mm

Fig. 7 Creep coefficient versus time for specimens at different distances from casting point

pulled-out under a sustain load progressively, while the β = [75–90°] specimens have a larger amount of fibres with a higher orientation angle towards the notch plane. In this case, the fibre reinforcement mechanism at the specimen's crack surface was mainly governed by matrix spalling at the fibres' exit points instead of fibre pulled-out. Therefore, the development of creep coefficient along time was influenced by the orientation of the crack plane within the SFRSCC panel, and this influence seems as larger as higher is the crack width.

3.5 *Influence of Distance from the Casting Point on the Creep Behaviour*

Figure 7 shows the influence of the specimens' distance to the casting point on the creep coefficient versus time relationship for specimens with F_a/F_{cr} = 100 %. It was opted to only present these relationships for the maximum F_a/F_{cr}, since they were considered as the most critical ones. Specimens positioned near the centre of the panel (i.e. casting point) presented a lower creep coefficient, while those located in the corner of the panel had the highest creep coefficients. This aspect could be ascribed to the decrease of the concrete flow velocity with the increase of the distance from the casting point. This decrease of the flow velocity influences both the fibre dispersion and orientation along the flow profile, and, consequently, leads to quite different fibre structures within the specimens at distinct distances from the casting point. From another point of view, this observation could be also ascribed to a decrease of the matrix strength with the increase of the distance from the casting point due to some segregation of the aggregate skeleton, resulting a weaker fibre/matrix interfacial bond strength that decreases the fibre reinforcement effectiveness.

3.6 Comparison Between Monotonic and Long-Term Behaviour

Figure 8 shows the comparison between the long-term assembled curves (according to the scheme presented in Fig. 3b) and the correspondent instantaneous monotonic responses, for $w_{cr} = 0.5$ mm and a high loading level, i.e. 75 % < $F_d/F_{cr} \leq 100$ %. Due to space limitation, the other series are not herein presented, but they can be found elsewhere [12]. These curves were obtained by averaging all responses. In general, it was concluded that the crack growth during the creep tests had a minor influence on the post-creep flexural behaviour. This remark can also be extended to the series not presented here. In fact, the assembled curves resemble quite well the average response from the monotonic tests. Nevertheless, in some cases, due to the results' scatter, as consequence of distinct fibre distributions, the assembled responses did not follow so closely the average monotonic curves, but nonetheless, they were yet comprised within the experimental envelope of the monotonic flexural tests.

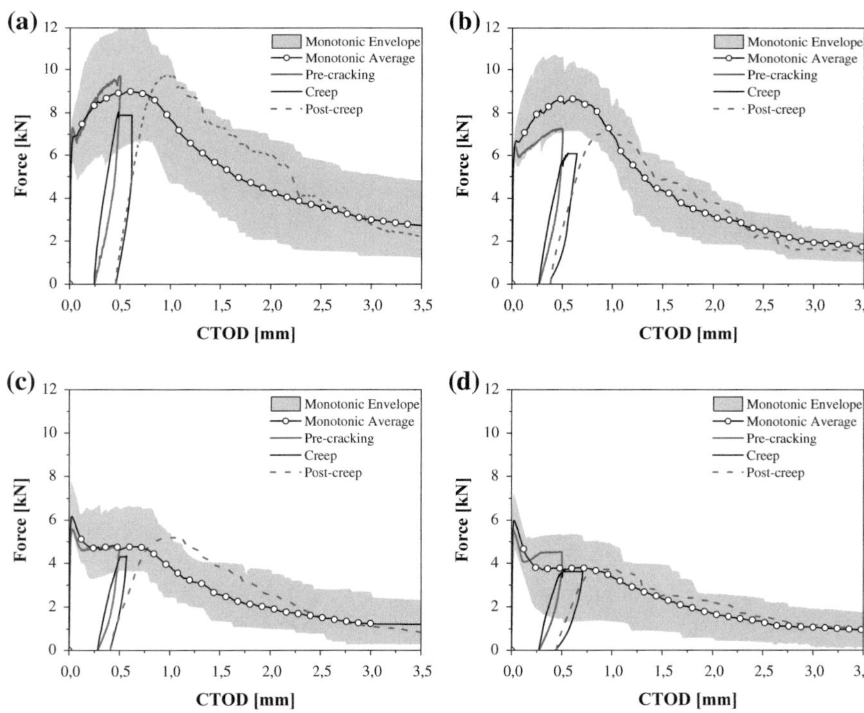

Fig. 8 Monotonic and long-term assembled curves for a pre-crack width of 0.5 mm and a high F_d/F_{cr} ratio: **a** β = [0–15°[; **b** β = [15–45°[; **c** β = [45–75°[; **d** β = [75–90°]

4 Analytical Equation to Predict the SFRSCC Creep Behaviour

In this section, an equation to predict the long-term response of the cracked SFRSCC is proposed. The influence of w_{cr} and F_a/F_{cr} parameters was taken into account. The following combined power and hyperbolic equation was used, since similar equations were already proposed to predict the creep behaviour of plain concrete, by e.g. [17]:

$$\varphi^C = t^A/(b+t^A) \tag{2}$$

where φ^C is creep coefficient, and t represents the time duration of loading (in hours). According to the experimental data, for each w_{cr}, the coefficient A is determined by:

$$A = w_{cr} \times (1-1/2F) + d \tag{3}$$

where w_{cr} is the pre-crack width (in mm) and F represents the level of loading (F_a/F_{cr}). In Eqs. (2) and (3) the constants b and d were determined by nonlinear curve fitting analysis procedures, and the following values were obtained: $b = 15$ ($R^2 = 0.94$) and $d = 0.17$ ($R^2 = 0.88$). It should be mentioned that Eq. (2) is only valid for $F_a/F_{cr} > 50$ %. Figure 9 shows the creep coefficient versus time obtained analytically and experimentally for each series. The proposed equation predicts with good accuracy the registered experimental data.

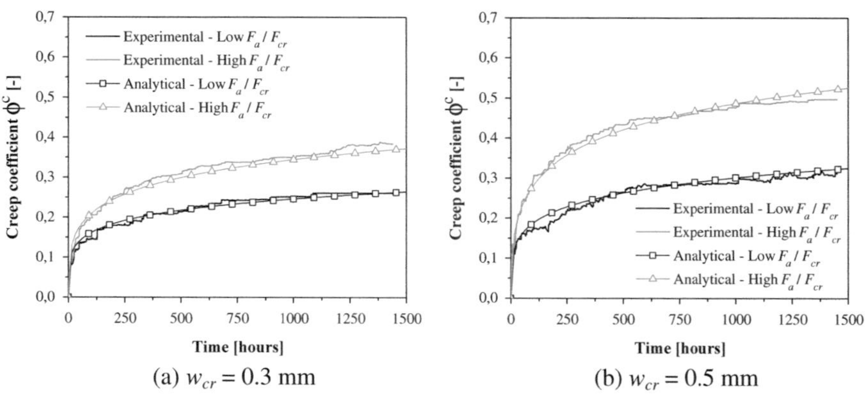

Fig. 9 Comparison between experimental and analytical creep coefficient versus time relationship

5 Conclusions

The present work reported the results of an experimental program to investigate the long-term behaviour of pre-cracked SFRSCC laminar structures of relatively small thickness. One hundred and twelve prismatic specimens were extracted from a SFRSCC panel. These specimens were notched with different orientations regarding to the expected SFRSCC flow direction, and were submitted to four-point flexural tests under instantaneous and sustained loads.

The long-term creep tests were performed for two pre-cracking levels and several loading levels. As expected, by increasing the level of the applied load, higher values of the creep coefficient were observed. Regarding the influence of the pre-cracking levels, w_{cr} = 0.5 mm series led to higher values of the creep coefficient than in the series w_{cr} = 0.3 mm, especially, if they were loaded with higher loading levels. Since w_{cr} = 0.5 mm was very close to the crack opening width correspondent to the one of the monotonic maximum load, the bond interface between fibre and matrix was more damaged, therefore the crack opening width under creep increased with a higher rate.

Specimens located nearer to the panel's corner showed a higher increase of creep coefficient with time. This aspect could be ascribed to the decrease of the concrete flow velocity with the increase of the distance from the casting point, which led to higher segregation of fibres and aggregates and lower strength paste with smaller effectiveness of the pull-out resisting mechanism of both fibres and aggregates.

The equation proposed to predict the creep coefficient for the developed SFRSCC when cracked up to 0.5 mm and loading levels in the interval $0.5 < F_a/F_{cr} \leq 1.0$ predicted with good accuracy the creep results.

Acknowledgments This work was supported by FEDER funds through COMPETE and National Funds through FCT, within ISISE unit, under the project *SlabSys*—FCOMP-01-0124-FEDER-020299, as well by the *Superconcrete* project w/ref. 645704 (H2020-MSCA-RISE-2014 funding program).

References

1. Balaguru, P.N., Shah, S.P.: Fiber Reinforced Cement Composites. McGraw-Hill Inc, New York (1992)
2. Barros, J.A.O., Figueiras, J.A.: Flexural behaviour of SFRC: Testing and modeling. ASCE Mater. Civ. Eng. **11**, 331–339 (1999)
3. Barros, J.A.O., Cunha, V.M.C.F., Ribeiro, A.F., Antunes, J.A.B.: Post-cracking behaviour of steel fibre reinforced concrete. Mater. Struct. **38**(1), 47–56 (2005)
4. Carmona, S., Aguado, A.: New model for the indirect determination of the tensile stress-strain curve of concrete by means of the Brazilian test. Mater. Struct. **45**, 1473–1485 (2012)
5. di Prisco, M., Ferrara, L., Lamperti, M.G.L.: Double edge wedge splitting (DEWS): an indirect tension test to identify post-cracking behaviour of fibre reinforced cementitious composites. Mater. Struct. **46**, 1893–1918 (2013)

6. Kurt, S., Balaguru, P.: Post-crack creep of polymeric fibre-reinforced concrete in flexure. Cem. Concr. Res. **30**, 183–190 (2000)
7. Oh, B.H., Park, D.G., Kim, J.C., Choi, Y.C.: Experimental and theoretical investigation on the post-cracking inelastic behaviour of synthetic fibre reinforced concrete beams. Cem. Concr. Res. **35**, 384–392 (2005)
8. Boshoff, W.B., Mechtcherine, V., van Zijl, G.P.A.G.: Characterizing the time-dependant behaviour on the single fibre level of SHCC: part 1: mechanism of fibre pull-out creep. Cem. Concr. Res. **39**, 779–786 (2009)
9. Arango, S.E., Serna, P., Marti-Vargs, J.R.: A test method to characterize flexural creep behaviour of pre-cracked FRC specimens. Exp. Mech. J. **52**, 1067–1078 (2012)
10. Zerbino, R.I., Barragan, B.E.: Long-term behaviour of cracked steel fibre-reinforced concrete beams under sustained loading. ACI Mater. J. **109**, 215–224 (2012)
11. Zhao, G., di Prisco, M.D., Vandewalle, L.: Experimental research on uni-axial tensile creep behaviour of pre-cracked steel fibre reinforced concrete. In: Fibre Reinforced Concrete: Challenges and Opportunities. Proceedings of the 8th RILEM International Symposium, Guimarães (2012)
12. Abrishambaf, A., Barros, J.A.O., Cunha, V.M.C.F.: Time-dependent flexural behaviour of cracked steel fibre reinforced self-compacting concrete panels. Cem. Concr. Res. **72**, 21–36 (2015)
13. Martinie, L., Roussel, N.: Simple tools for fiber orientation prediction in industrial practice. Cem. Concr. Res. **41**, 993–1000 (2001)
14. Boulekbache, B., Hamrat, M., Chemrouk, M., Amziane, S.: Flowability of fibre-reinforced concrete and its effect on the mechanical properties of the material. Constr. Build. Mater. **24**, 1664–1671 (2010)
15. Abrishambaf, A., Barros, J.A.O., Cunha, V.M.C.F.: Relation between fibre distribution and post-cracking behaviour in steel fibre reinforced self-compacting concrete panels. Cem. Concr. Res. **51**, 57–66 (2013)
16. UNI11039: Steel fibre reinforced concrete, Part I: Definition, classification, specification and conformity, Part II: Test method for measuring first crack strength and ductility indexes. Italian Board for Standardization (2005)
17. CEB-FIP: Volume 1, Model Code 2010. Tomas Telford, Lausanne, Switzerland (2012)
18. EFNARC: The European guidelines for self-compacting concrete (2005)
19. Cunha, V.M.C.F., Barros, J.A.O., Sena-Cruz, J.M.: An integrated approach for modeling the tensile behaviour of steel fibre reinforced self-compacting concrete. Cem. Concr. Res. **41**(1), 64–76 (2011)
20. EN1992-1-1: Eurocode 2: Design of Concrete Structures. European Committee of standardization, Brussels (2004)

Part IV
Creep Testing Methods

Effect of Residual Strength Parameters on FRC Flexural Creep: Multivariate Analysis

Emilio Garcia-Taengua, Aitor Llano-Torre, Jose R. Marti-Vargas and Pedro Serna

Abstract This paper reports the multivariate analysis of experimental results from more than one hundred FRC prismatic specimens tested under sustained flexural loads for at least 90 days, collected from previously published sources. Principal Component Analysis was used to minimise the number of variables in the modelling process while compromising the minimum amount of information. The creep parameters analysed were the creep coefficients at 14, 30, and 90 days and the corresponding crack opening rates. They were related to the following factors: concrete compressive strength, residual load-bearing capacity in flexure, fibre material, and load. Multiple linear regression was used for the modelling of these relationships. Higher levels of flexural toughness were found to significantly reduce the variability of all creep parameters. Differences in fibre material were detected to introduce important differences in interaction with other factors, especially the load ratio, which was attributed to the association between the range of flexural toughness to be expected and the type of fibres used.

Keywords Fibre material · Flexure · Linear models · Multivariate · Residual strength · Toughness

1 Introduction

The major feature of fibre reinforced concrete (FRC) is its residual load-bearing capacity when cracked, in terms of flexural residual strength and flexural toughness [1].

E. Garcia-Taengua (✉)
School of Civil Engineering, University of Leeds, Leeds LS2 9JT, UK
e-mail: e.garcia-taengua@leeds.ac.uk

A. Llano-Torre · J.R. Marti-Vargas · P. Serna
ICITECH Institute for Concrete Science and Technology, Universitat Politècnica de València, Camino de vera s/n Building 4N, 46022 València, Spain

P. Serna et al. (eds.), *Creep Behaviour in Cracked Sections of Fibre Reinforced Concrete*, RILEM Bookseries 14,
DOI 10.1007/978-94-024-1001-3_12

FRC show considerable scatter in their flexural response [2], and in consequence their deformational behaviour when under sustained flexural loads shows high levels of variability, especially in its cracked state. A number of different factors affect creep of cracked FRC elements, relative to the composition and mechanical properties of the concrete matrix, fibres type and dosage, environmental conditions, loading conditions, or time, among other factors [3, 4].

Previously published papers and reports concerned with the flexural response of cracked FRC elements under sustained loads contain important volumes of experimental results obtained from different FRC mixes produced with different types of fibres at different dosages. The authors of this paper have been working on the compilation of analysis of these experimental data [5]. This information gathered from different sources can be collectively analysed using a data mining approach in an attempt to draw conclusions that might be generalised beyond the particular limitations of any individual study.

2 Overview of Data and Methodology

2.1 Summary of the Information Collected

This research was based on the creep test results corresponding to 118 FRC specimens, tested under sustained flexural loads for at least 90 days. For each of these cases, the information collected included parameters concerning the following aspects:

- Concrete matrix: maximum aggregate size, average compressive strength at 28 days (f_c, MPa).
- Fibres employed: material (glass, steel or synthetic), length, aspect ratio, and fibres content as volume fraction.
- Flexural residual strength: limit of proportionality f_L, and residual strength parameters f_{R1}, f_{R2}, f_{R3} and f_{R4} (MPa) as per the standard EN 14651 [6].
- Sustained load during the creep test typified by the residual capacity, or load ratio IFa [7].
- Creep coefficients at the ages of 14, 30, and 90 days [7].
- Crack opening rates (CORs) between 0 and 14 days, 14 and 30, and 30 and 90 days (COR^{0-14}, COR^{14-30}, and COR^{30-90} respectively) [7].

Representative values for each of the variables are given in Table 1 for reference. Minimum, average and maximum values were intended to be descriptive of the entire set of data and were calculated as the 5, 50, and 95 % percentiles. The percentage of missing values for each variable represents those cases for which complete information could not be obtained.

Table 1 Summary of the information in the database

	Minimum (5 %-percentile)	Average	Maximum (95 %-percentile)	Percentage of missing values (%)
Fibres content, V_f (%)	0.47	0.58	0.95	0.0
Fibre material	n.a.	n.a.	n.a.	0.0
Fibre length (mm)	35.00	42.89	54.00	0.0
Fibre slenderness	45.00	71.09	158.00	0.0
Max. aggregate size (mm)	10.00	12.46	20.00	11.0
Compr. strength f_c (MPa)	34.00	42.31	56.54	0.0
Limit of proportionality f_L (MPa)	2.97	4.32	6.04	23.7
Residual strength f_{R1} (MPa)	0.87	3.65	9.55	0.8
Residual strength f_{R2} (MPa)	0.79	3.92	9.06	36.4
Residual strength f_{R3} (MPa)	0.78	4.01	10.09	32.2
Residual strength f_{R4} (MPa)	0.60	3.80	9.57	31.3
Load ratio IFa (%)	36.90	66.21	89.56	0.0
Creep coefficient at 14 days	0.09	0.70	1.93	0.8
Creep coefficient at 30 days	0.11	0.91	2.19	0.0
Creep coefficient at 90 days	0.31	1.26	3.11	2.5
COR between 0–14 days	0.86	15.11	41.07	0.0
COR between 14–30 days	0.20	3.58	10.15	0.0
COR between 30–90 days	0.27	1.51	4.16	0.0

2.2 *Treatment of Missing Values*

As the sources of these data did not always report the mix characteristics, their flexural response or their flexural creep behaviour in the same way, some of the values of interest had not been reported.

This is quite common in the context of data mining applications or multivariate statistics [8], and different strategies can be adopted to moderate the impact of missing cases on the reliability of the information that can be extracted from a set of data. In this case, the prevalence of missing values was relatively small and did not justify complete elimination of any cases. Instead, a multiple imputation by fully conditional specification [9] was done to simulate the missing values taking advantage of the multiple correlations between the different variables.

3 Multivariate Description of FRC Mechanical Performance

Six variables were considered as descriptors of the mechanical performance of each FRC mix: f_c, f_L, f_{R1}, f_{R2}, f_{R3} and f_{R4}, all of them expressed in MPa.

After some preliminary analyses, it was observed that this number of variables needed to be reduced in order to correlate the material's performance with the flexural creep parameters and the level of sustained load applied. A multivariate analysis was carried out in an effort to find a compromise between the reduction of the number of variables and the amount of information unaccounted for.

3.1 Bivariate Correlations

The Pearson correlation coefficients [10] between any pair of these six variables are shown in Table 2. Very strong correlations were observed between the residual flexural strength parameters (values between 0.902 and 0.978). On the other hand, the average compressive strength of concrete was practically uncorrelated to these parameters. The relationship between any of these and the limit of proportionality was less clear, as it was found to be slightly correlated to the compressive strength and moderately to the residual flexural strength parameters.

3.2 Principal Component Analysis (PCA)

The structure of correlations in Table 2 proved that these six variables are not independent and therefore they could not be treated as such when modelling their effect on creep. Principal Component Analysis (PCA) was used to condense the information they describe into a reduced set of variables.

After these variables were centered and scaled to unit variance, principal components were extracted by singular value decomposition of the correlation matrix [11], and a Varimax rotation was applied [12]. The first three principal components were retained as sufficiently informative, explaining 97.11 % of the total variance in the original variables.

Each one of the principal components is a linear combination of the original variables, and therefore they define new rotated axes PC1, PC2 and PC3. By plotting the weights of the original variables in these linear combinations against the new coordinate system defined by PC1, PC2, and PC3, the two biplots in Fig. 1 were obtained.

Table 2 Correlation matrix

	f_c	f_L	f_{R1}	f_{R2}	f_{R3}	f_{R4}
f_c	(1.000)	0.310	0.055	0.090	0.066	0.076
f_L		(1.000)	0.654	0.633	0.598	0.572
f_{R1}			(1.000)	0.954	0.939	0.908
f_{R2}				(1.000)	0.949	0.902
f_{R3}					(1.000)	0.978
f_{R4}						(1.000)

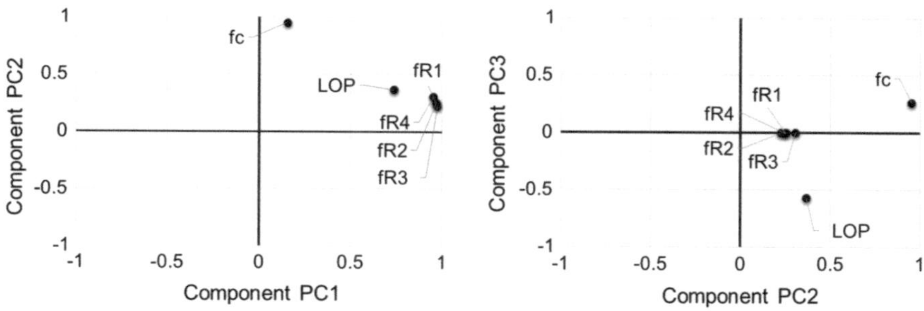

Fig. 1 Biplots for interpretation of the principal components after the PCA

The four residual flexural strength parameters formed a very clear cluster. This cluster defined the first component PC1, and therefore PC1 was interpreted as representative of the residual load-bearing capacity. Furthermore, as the distances between f_{R1}, f_{R2}, f_{R3} and f_{R4} were negligible, it followed that PC1 values were proportional to $f_{R1} + f_{R2} + f_{R3} + f_{R4}$,

Concrete compressive strength, on the other hand, was clearly apart from this cluster, consistently with what the bivariate correlations indicated. Therefore, the second component PC2 was defined by compressive strength alone.

These two components PC1 and PC2 described 89.70 % of the total variance. The third component (PC3), accounted for only 7.41 %, and had no clear interpretation, as neither the limit of proportionality nor any of the other variables were found to have a determining weight.

3.3 Redefinition of the Variables Characterising FRC Mixes

The results of the PCA clearly indicate that the mechanical performance of FRC mixes as described by the six initial variables can be effectively reduced to only two variables while retaining nearly 90 % of the total variance:

- Concrete compressive strength, f_c, in MPa.
- Toughness, defined as $T = f_{R1} + f_{R2} + f_{R3} + f_{R4}$, in MPa.

4 Multiple Regression Analysis

4.1 Definition of Regressors

As concluded in the previous section, two variables were retained as representative of the mechanical performance of the material: the average compressive strength at 28 days (f_c, in MPa), and the flexural toughness (T, in MPa).

The load ratio (*IFa*, in percentage) was considered to account for the sustained load applied.

The fibre material was also considered, as a binary variable (*Mat*): synthetic or non-synthetic, as synthetic fibres were found to be significantly different (*p*-value = 0.00125).

4.2 Summary of the MLR Models Obtained

Multiple linear regression (MLR) was used to relate each of the creep parameters under consideration to the aforementioned regressors. Initially, models with different formulations including all variables, squared variables, second order interactions and some selected third order interactions were considered. Stepwise and best subsets regression procedures were applied to simplify these models by discarding those terms that were not statistically significant.

Table 3 summarises the MLR models as obtained after this sequential process. The terms that were identified as statistically significant (*p*-values up to 0.05) are marked with an asterisk. These models were used as a tool to obtain average

Table 3 Summary of the MLR models

	Creep coefficients			Crack opening ratios		
	14 days	30 days	90 days	0-14 days	14-30 days	30-90 days
IFa						*
IFa^2						*
f_c						
T					*	*
Mat	*	*	*			
$T \times IFa$	*					
$T \times fc$	*	*	*			*
$Mat \times IFa$	*	*	*	*	*	*
$Mat \times IFa^2$	*	*	*	*	*	*
$Mat \times T$	*	*	*			
$Mat \times f_c$			*	*	*	*
$Mat \times f_c^2$	*	*	*	*	*	*
$Mat \times IFa \times T$	*	*	*	*		
$Mat \times IFa \times f_c$	*	*	*	*	*	*
$Mat \times IFa^2 \times T$	*	*		*		
$Mat \times IFa^2 \times f_c$	*	*	*	*	*	*
R-squared	0.55	0.51	0.46	0.54	0.43	0.41

estimates and trends for the creep coefficients, which allowed the study of these multivariate relationships as reported in the following sections. Furthermore, it is anticipated that future post-processing of these models can increase their predictive accuracy.

5 Discussion of the Models for Creep Coefficients

The MLR models summarised in Table 3 made it possible to plot response curves for the creep coefficients with respect to load ratio, concrete compressive strength and toughness. Furthermore, the differences introduced by the fibre material and their modifying effect on other variables were also scrutinised.

These aspects were examined through the contrastive effects plots for the MLR models obtained, using the open source package "visreg" in R [13]. In these plots the relative change in the creep coefficients is represented, instead of their predicted average values. This way the effect of any one variable could be analysed without assuming constant values for the other variables, which is an important advantage when interpreting complex models with many interactions.

5.1 *Effect of Concrete Compressive Strength*

Variations in the creep coefficients with respect to concrete compressive strength are shown in Fig. 2. Average trends as obtained from the MLR models are drawn with continuous lines, while dots correspond to individual data. Coloured bands represent the 90 %-confidence band.

It was observed that higher compressive strength values tend to reduce the creep coefficients at 30 and 90 days, following a quadratic trend. However, this relationship between concrete compressive strength and changes in the creep coefficient was not detected to be significant at 14 days.

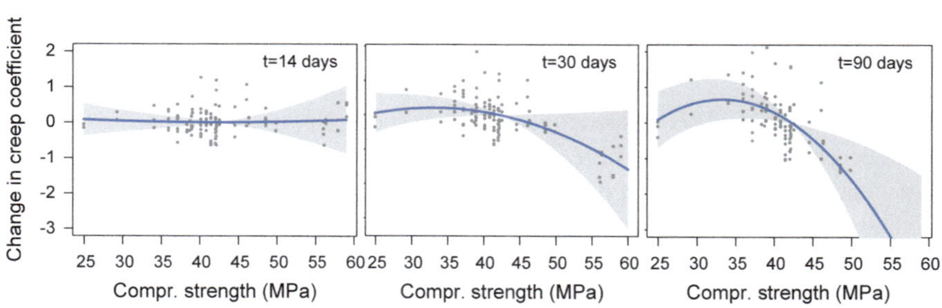

Fig. 2 Effect of compressive strength on creep coefficients

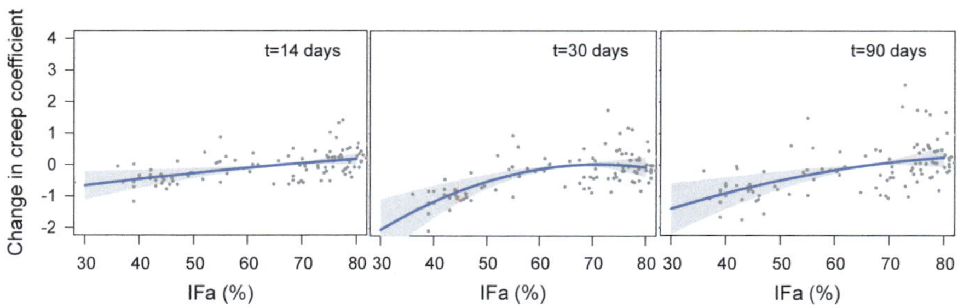

Fig. 3 Effect of load ratio on creep coefficients

5.2 Effect of Loading Ratio and Fibre Material

Figure 3 shows the variation in the creep coefficients at 14, 30, and 90 days with respect to the load ratio applied during the creep tests. Higher load ratios implied higher creep coefficients at all ages. However, the effect of increasing the load ratio was less noticeable on the creep coefficient at 14 days: an increase of 10 % in *IFa* led to an average increase of the creep coefficient in 0.2, whilst this increase was of 0.35 at 90 days.

These average trends were not significantly affected by the fibres length or aspect ratio. However, interesting differences were observed when the interaction between fibre material and load ratio was analysed in addition to the simple effect of increasing the load ratio. Figure 4 shows the contrastive effects plots with respect to the load ratio but distinguishing between synthetic and non-synthetic fibres.

At lower load ratios, creep coefficients were observed to be systematically higher in those cases where synthetic fibres had been used. The gap between them, however, was gradually reduced for increasing load ratios, and differences due to the fibre material were negligible for load ratios of 60 % or higher. This pattern was consistent for all three ages considered.

However, the trends represented in Fig. 4 can be misleading: it is important to emphasise that Fig. 4 shows an association between the fibre material and higher creep coefficients, but this is not necessarily of a cause-effect nature. In fact, the analysis of variance on the squared residuals of these MLR models revealed that only one variable had a statistically significant effect on the variability of creep coefficients, and therefore on the differences represented in Fig. 4. And it was not the fibre material, but the flexural toughness T (p-values = 0.056, 0.020, and 0.016 at 14, 30, and 90 days respectively). In consequence, the differences between synthetic and non-synthetic fibres in Fig. 4 could not be directly attributed to the fibre material: the role played by the flexural toughness in these models and its interaction with the fibre material needed to be explored more closely, and that is precisely what is reported in the following section.

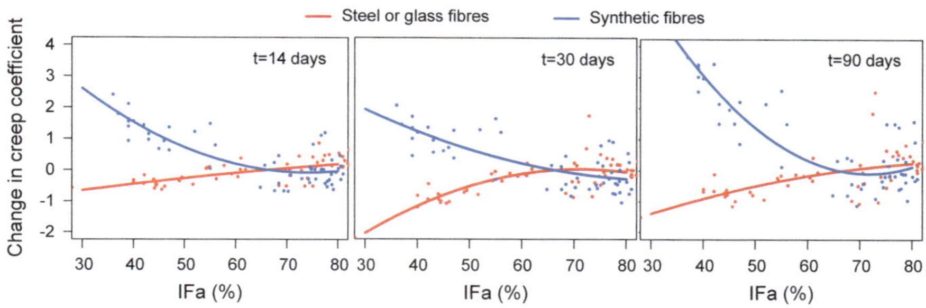

Fig. 4 Effects of fibre material and load ratio on creep coefficients

5.3 *Effect of the Flexural Toughness*

The effect of increasing flexural toughness on creep coefficients at 14, 30 and 90 days is shown in Fig. 5. The same average trend was consistently observed at all ages: creep coefficients are reduced when the flexural toughness is increased, although this was less pronounced at the age of 14 days. However, it was observed that creep coefficients presented increasing variability with age, especially so for the lower toughness values, as indicated by the scatter of the dots in Fig. 5. Enhancing the flexural toughness has therefore the advantage of reduced creep coefficient, with more stable and predictable values.

When the distinction between synthetic and non-synthetic fibres was accounted for, Fig. 6 was obtained. It was observed that, as long as a similar level of flexural toughness is achieved, the choice between synthetic or non-synthetic fibres does not introduce a substantial difference in terms of creep coefficients. Therefore, Fig. 6 confirms that the differences observed in Fig. 4 cannot be directly attributed to the fibre material, because generally synthetic fibres are used in dosages that lead to considerably lower levels of flexural toughness when compared to elements with steel fibres. If this aspect is disregarded, cause-effect relationships can be misinterpreted, introducing a bias in favour of the steel fibres not sufficiently supported by evidence.

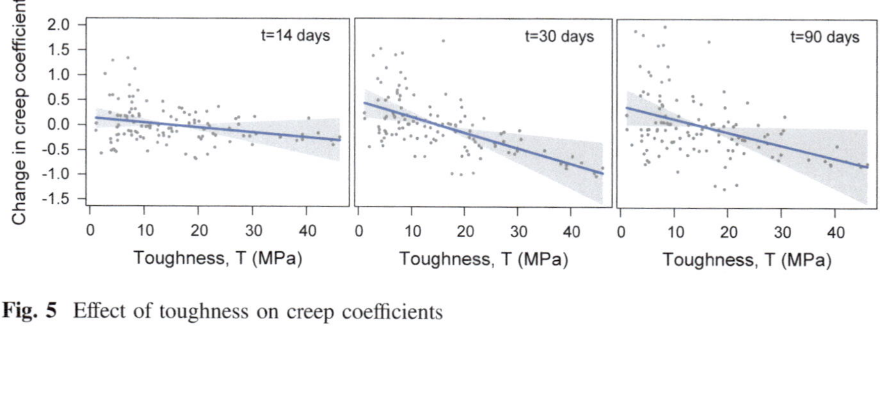

Fig. 5 Effect of toughness on creep coefficients

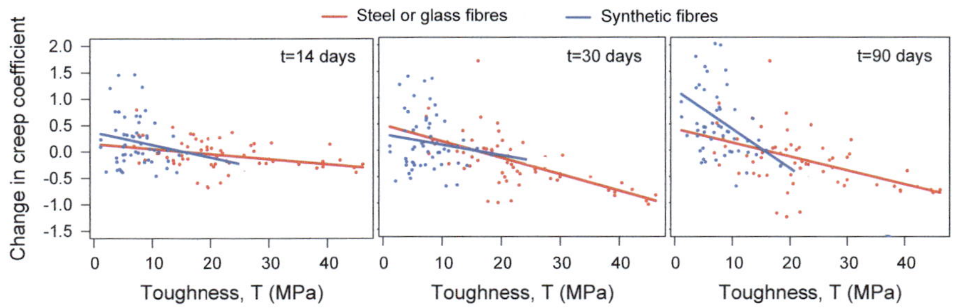

Fig. 6 Interaction between fibre material and toughness on creep coefficients

With respect to the variability of the creep coefficients, both Figs. 5 and 6 confirm that their values are more scattered when lower levels of flexural toughness are considered.

6 Discussion of the Models for the Crack Opening Rates

6.1 *Effect of Load Ratio and Fibre Material*

Figure 7 shows the effect of increasing the load ratio on the variation of crack opening rates for the three timespans considered, distinguishing between synthetic and non-synthetic fibres. The relationship between the crack opening rates and the load ratio was not significantly modified by the fibre material. Although more variability was detected for COR^{0-14}, this was significantly reduced afterwards and the increasing trend with respect to the load ratio was gradually moderated.

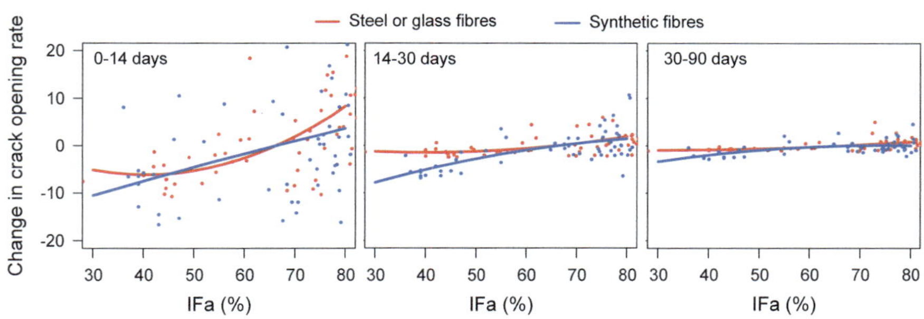

Fig. 7 Effect of load ratio on the crack opening rates

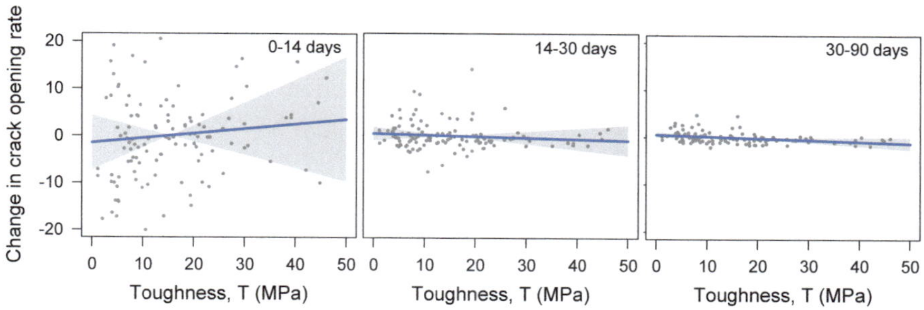

Fig. 8 Effect of toughness on the crack opening rates

6.2 Effect of Flexural Toughness

In consistency with the findings reported for the creep coefficients, the effect of fibre material on the crack opening rates was examined in conjunction with that of flexural toughness. Figure 8 shows the average trends for the crack opening rates with respect to the flexural toughness. In terms of relative magnitude, the effect of flexural toughness on crack opening rates was more moderate than that of load ratio if Figs. 7 and 8 are compared. It is also interesting to note that, if Figs. 5 and 8 are compared, the effect of increasing flexural toughness on creep coefficients was more noticeable than on the crack opening rates.

COR^{14-30} and COR^{30-90} were detected to slightly decrease in average when flexural toughness values increased. On the other hand, COR^{0-14} showed a considerably higher variability, and this was one of the reasons why the first effects plot in Fig. 8 seems inconsistent with the other two.

Such inconsistency was resolved when the distinction between cases with synthetic fibres and non-synthetic fibres was considered, as shown in Fig. 9. The fibre material introduced no substantial modification to the relationship between the flexural toughness and the crack opening rates after 14 days. However, with respect to COR^{0-14}, it was evidenced that data corresponding to synthetic fibres followed a

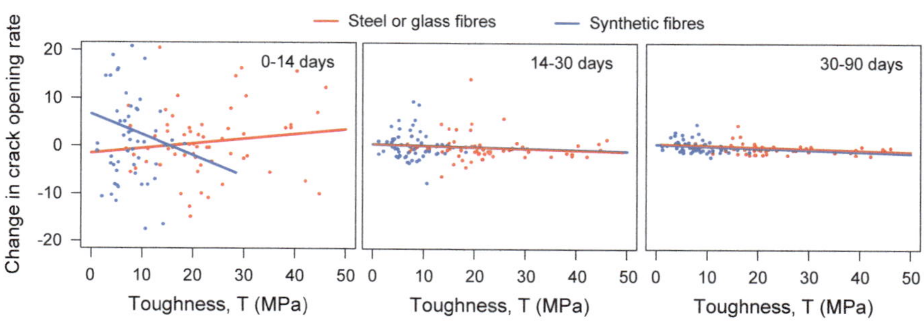

Fig. 9 Effect of fibre material and toughness on the crack opening rates

different trend, thus the apparently excessive variability noticed in Fig. 8. When the fibres used are synthetic, improvements in flexural toughness clearly reduce COR^{0-14}. After that age, the interaction between the fibre material and the flexural toughness is not significant in terms of crack opening rates.

7 Conclusions

In this paper, a set of experimental results corresponding to more than one hundred FRC prismatic specimens tested under sustained flexural loads were analysed by means of Principal Components Analysis and Multiple Linear Regression. The following conclusions were drawn:

- The application of multivariate semi-empirical modelling techniques is a useful approach to advance in the understanding of flexural creep of cracked FRC elements. The collection of a representative number of data from different mix compositions, types of fibres, and subject to different sustained load levels will feed more comprehensive meta-analyses in the future.
- Three parameters have been detected to have a determining impact on the flexural creep response of cracked FRC sections: concrete compressive strength, the load ratio, and the flexural toughness defined as the sum of the residual flexural strength values f_{R1}, f_{R2}, f_{R3} and f_{R4}.
- Differences in fibre material do not play a direct determining role on the response of cracked FRC sections under sustained flexural loads. Rather, their influence is on the flexural toughness of the material, which in turn affects the creep response.
- A general trend towards stabilisation in time is observed and consistent for all the creep coefficients and crack opening rates analysed, even in those cases when the load ratio is high.

References

1. Di Prisco, M., Plizzari, G., Vandewalle, L.: Fibre reinforced concrete: new design perspectives. Mater. Struct. **42**(9), 1261–1281 (2009)
2. Cavalaro, S., Aguado, A.: Intrinsic scatter of FRC: an alternative philosophy to estimate characteristic values. Mater. Struct. **48**(11), 3537–3555 (2015)
3. Vasanelli, E., Micelli, F., Aiello, M.A., Plizzari, G.: Long term behavior of FRC flexural beams under sustained load. Eng. Struct. **56**, 1858–1867 (2013)
4. Garcia-Taengua, E., Arango, S., Marti-Vargas, J.R., Serna, P.: Flexural creep of steel fiber reinforced concrete in the cracked state. Constr. Build. Mats. **65**, 321–329 (2014)
5. Llano-Torre, A., Garcia-Taengua, E., Marti-Vargas, J.R., Serna, P.: Compilation and study of a database of tests and results on flexural creep behavior of fibre reinforced concrete specimens. In: Proceedings of FIB Symposium Concrete Innovation and Design, Copenhagen, 2015

6. European Standard EN 14651:2005+A1:2007: Test Method for Metallic Fibre Concrete—Measuring the Flexural Tensile Strength (Limit of Proportionality (LOP), Residual). European Committee for Standardization (2007)
7. Arango, S., Serna, P., Martí-Vargas, J.R., García-Taengua, E.: A test method to characterize flexural creep behaviour of pre-cracked FRC specimens. Exp. Mech. **52**, 1067–1078 (2012)
8. Lakshminarayan, K., Harp, S., Samad, T.: Imputation of missing data in industrial databases. Appl. Intellig. **11**, 259–275 (1999)
9. Lee, K.J., Carlin, J.B.: Multiple imputation for missing data: fully conditional specification versus multivariate normal imputation. Am. J. Epidem. **171**, 624–632 (2010)
10. Rodgers, J. L., Nicewander, W. A.: Thirteen ways to look at the correlation coefficient. Am. Stat. **42**(1), 59–66 (1988). doi:10.2307/2685263
11. Wold, S., Esbensen, K., Geladi, P.: Principal component analysis. Chemom. Intell. Lab. Syst. **2**, 37–52 (1987)
12. Browne, M.W.: An overview of analytic rotation in exploratory factor analysis. Multivar. Behav. Res. **36**, 111–150 (2011)
13. Breheny, P., Burchett, W.: visreg: Visualization of regression models. R package version 11–1 (2012). Online documentation: https://CRAN.R-project.org/package=visreg

Mid-term Behaviour of Fibre Reinforced Sprayed Concrete Submitted To Flexural Loading

Catherine Larive, Damien Rogat, David Chamoley, André Regnard, Thibaut Pannetier and Christine Thuaud

Abstract Creep is a major research issue in fibre reinforced concrete. Many test methods exist to assess creep on structural elements subjected to mid or long-term flexural loading. Most of them are performed on notched beams on isostatic support. Asquapro launched an experimental program to assess the creep behaviour of fibre reinforced sprayed concrete (FRSC). A test method was especially designed to take into account the specifications of FRSC for underground support (unnotched specimens under statistically indeterminate support conditions). The aim of this paper is to present the results of this programme after one year of testing. Differences between fibres are clearly identified, within the same type of fibres or between different types (metallic/polymeric). A comparison is also made with specimens reinforced by a wire welded mesh.

Keyword Concrete

1 Introduction

The expression 'Mid-term behaviour' used in the title corresponds to creep tests lasting one year.

Creep is one of the most controversial issues concerning fibre reinforced concrete. RILEM has recently devoted a working group to this subject in order to

---

C. Larive (✉) · D. Chamoley · T. Pannetier
CETU, Bron, France
e-mail: catherine.larive@developpement-durable.gouv.fr

D. Rogat
Sigma Béton, L'Isle-d'Abeau, France

A. Regnard
Asquapro, Bron, France

C. Thuaud
SNCF Réseau, La Plaine Saint Denis, France

P. Serna et al. (eds.), *Creep Behaviour in Cracked Sections of Fibre Reinforced Concrete*, RILEM Bookseries 14,
DOI 10.1007/978-94-024-1001-3_13

better understand this phenomenon. Some results are already available but most of them are related to cast concrete.

Static conditions of concrete elements are one of the main differences between FRC and FRSC structures. Indeed, in underground work, FRSC is always under statically indeterminate conditions because of the surrounding rock support. Consequently, a redistribution of the loads is essential after cracking to ensure that the sprayed concrete ring is able to support rock pressure and allow reasonable displacements as described in both the NATM and the convergence-confinement analysis.

This difference requires a specific test method to evaluate the creep behaviour of FRSC. Asquapro, the French Association for the Quality of Sprayed Concrete, decided to elaborate such a specific test method and to launch an experimental programme on that subject. The first part of this paper describes the test method, the second one the experimental programme and the last one the main results.

2 Test Method to Evaluate the Mid-term Behaviour of Frsc Submitted to Flexural Loading

The temperature and hygrometry conditions of the room are measured throughout the tests and must be consistent if several tests are performed in the same room.

The creep test is performed on three slabs of 60 × 60 cm with a thickness of 10 cm, as in the NF EN 14488-5 test method to measure the energy absorption capacity. Three other slabs made of the same concrete matrix but without fibres are necessary to determine the appropriate loading.[1] Four slabs of each concrete must be sprayed in case a problem arises with one of them.

The test requires a creep test device (Fig. 1) designed and used as follows:

Stage 1: characterisation of the reference concrete (BB)
Seven days after spraying, three unreinforced slabs are loaded with a press according to the NF EN 14488-5 test procedure until the maximum force is obtained. The mean value is called $F_{el\text{-}max\ BB}$ and is used to determine the loads that shall be applied to the reinforced slabs thereafter.
Stage 2: calibration of the creep test device
During the creep test, one or two loads may be applied on the slab through a square loading block of 10 × 10 cm, depending on the measured deflection: the first one is $F_{pre\text{-}cracking} = 1.2 * F_{el\text{-}max\ BB}$, the second one is $F_{creep} = 0.5\ F_{pre\text{-}cracking}$. The load is reduced to F_{creep} only if the deflection exceeds 2 mm.
Calibration is carried out using a load cell. A steel slab is placed in the same location as the concrete slabs to be tested; centring is carried out automatically by clamping guides.

[1]This concrete is considered as the reference concrete (BB).

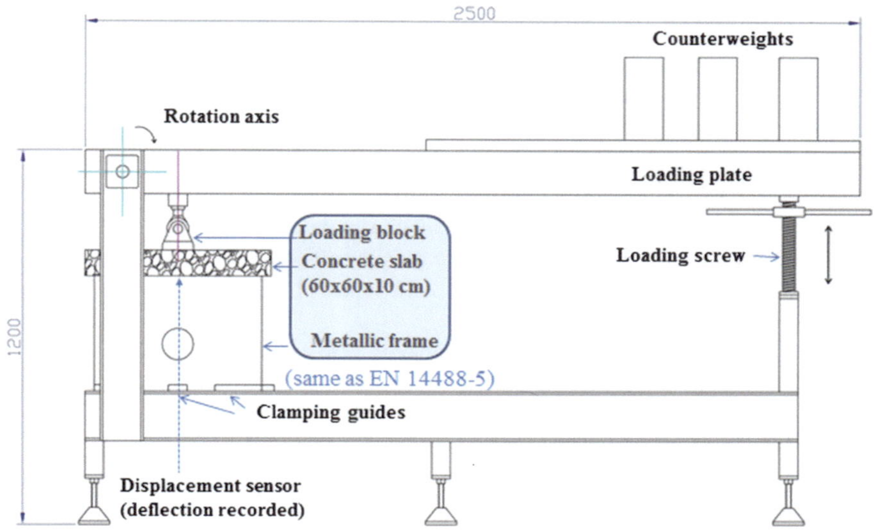

Fig. 1 Creep test device

The counterweights are determined firstly to obtain the lowest value: F_{creep}. The corresponding weights are then marked. They must remain in place during the whole test. Then the counterweights to be added to obtain the highest load, $F_{pre\text{-}cracking}$, are determined.

Stage 3: pre-cracking of a reinforced slab and beginning of the deflection measurement

Prior to cracking, both diagonals are drawn on the underside of each slab (in contact with the mould). A square of 1 × 1 cm is marked out in the middle, and then the slab is centred on a metallic frame on which it will stay until the end of the test.

Pre-cracking is carried out seven days after spraying, in two separate phases:

Phase 1: the slab is loaded with a testing machine according to the NF EN 14488-5 test method, until the first drop in the force is detected; then the machine is immediately stopped. As this drop is brutal, it has been found more efficient to rely on the ability of an experienced technician rather than trying to detect the load peak automatically.

Phase 2: the slab, still on its frame, is positioned on the creep test device (the slab and the frame are transported together from the press to the creep device). If necessary the slab is re-centred on its frame, which is also centred within the device, by using clamping guides. A LVDT-type displacement sensor is installed within the central square that was marked out, making sure that it is not positioned over any cracks.

The pre-cracking load $F_{pre\text{-}cracking}$ is applied by unscrewing the screw positioned underneath the loading plate. The recording of deflection values begins from the start of loading. The load is maintained constant for 1 year, except if the deflection exceeds 2 mm.

Stage 4: reduction of the load and continuous measurement of deflections (if necessary)
If a deflection of 2 mm is measured, the load is reduced to the value of $F_{creep} = 0.5 * F_{pre\text{-}cracking}$. The deflection is continuously monitored.
Stage 5: load removal and measurement of recovery deflection
After one year of testing, the load is removed from the slab. The recording of deflection continues for one week after load removal. The slab is then removed from the creep test device.
Stage 6: measurement of the residual energy absorption capacity
The slab is tested according to the NF EN 14488-5 test to evaluate its residual energy absorption capacity.
Stage 7: checking of the calibration of the creep test device
The load value of the creep test device is checked using the same equipment as for the calibration (the metal plate and load cell).

Stages 6 and 7 can be carried out in any order. Photos and additional information on the elaboration of this test method are available in [1].

3 Experimental Programme

The experimental campaign organised by Asquapro had two main aims: (1) To study the contribution of different fibres to the behaviour of sprayed concrete subjected to punching/bending stresses; (2) To develop a "standardisable" test for studying the creep behaviour of FRSC under a constant load.

3.1 Testing Programme

The programme consisted in characterising the fresh concrete properties and measuring the compressive strength, the energy absorption capacity and the creep deformation. More than 100 specimens were necessary to carry out the programme.

3.2 Concrete Mix Design

To ensure good comparability, the same concrete matrix was used for all types of concrete: reinforced or plain. A typical mix design used in France for sprayed concrete for underground support was chosen (see Table 1). Only the dosage of

Table 1 Concrete mix design

Component	Designation	Dosage (kg/m^3)
Sand	0–4 mm	1.041
Gravel	4–11.2 mm	729
Cement	CEM I 52.5N SR3 CE PM-CP2 NF	400
Admixture	Tempo 11 (0.2–0.4 %)	0.8–1.6
Water	W/C = 0.475	190

superplasticiser was slightly modified from one type of concrete to another according to the different fibres. The grading envelope is consistent with the Asquapro recommendation and the NF 95102 French standard.

3.3 Concrete Reinforcements

Six different fibres and a welded wire mesh were used to reinforce the sprayed concrete elements. The fibre dosage was proposed by each supplier, in compliance with the mixes commonly used in concrete for underground support (see Table 2). The welded mesh is the one used by the French National Railway Company in operations to reinforce existing tunnels. The reinforcement characteristics of the Reinforced Sprayed Concretes (RSC) were as follows.

3.4 Application

The spraying was carried out by a single operator using a Meyco Potenza robot. The dosage of the activator (Sika AF53) was electronically controlled by the concrete flow rate and set at a single value of 7 % for all of the concrete test specimens.

Table 2 Reinforcements characteristics

RSC	Type of reinforcement	Fibre dosage (kg/m^3) or welded mesh characteristics	E (GPa)	Tensile strength (MPa)
BF01	Polymeric fibres	8	6	550
BF02	Steel fibres	30	210	1650
BF03	Polymeric fibres	7	12	580
BF04	Steel fibres	30	210	1345
BF05[a]	Polymeric fibres	7	4.8	470
BF06	Amorphous metallic fibres	25	130	1800
WWM	Welded wire mesh	ST 50, 30 × 10 cm^2 mesh, 8 mm diameter	210	540

[a]This is a prototype fibre, not on the market

Table 3 Compressive strength (MPa)

Compressive strength at 1, 7 and 28 days	BB	BF01	BF02	BF03	BF04	BF05	BF06	WWM
Day 7 (MPa)	36.7	35.4	36.3	32.6	32	37.5	35.7	36.7
Day 28 (MPa)	44.1	45.8	49.2	47.2	45.1	47.5	48.8	44.1

3.5 Basic Characteristics

Concrete slump, fresh concrete densities, air contents and compressive strength (see Table 3) were measured. They were very similar for all the samples. Slump measurements varied from 180 to 230 mm. Density varied from 2.360 to 2.375 kg/m^3 and the air content from 1.5 to 2.1 %. The consistency of the results confirmed that the production and application were properly controlled.

Compressive strengths were measured, according to the EN 12390-3 test method, on cores (diameter 64, 128 mm high) extracted from sprayed panels. Mean compressive strength at 28 days was 46.5 MPa.

4 Main Results

4.1 Consistency of the Experimental Loads

The following values of $F_{el\text{-}maxBB}$ were measured: 31.78; 29.61; 37.69 kN (mean value: $F_{el\text{-}max\ BB}$ = 33.03 kN). Thus, calculations lead to: $F_{pre\text{-}cracking}$ = 39.63 kN and F_{creep} = 19.81 kN.

A comparative study was carried out to check if these loading rates are consistent with what could be observed in a tunnel project. To do so, structural analysis using Navier's solution and a yield line analysis were carried out on the slab used in the creep test. They led to mean bending moments of $M_{pre\text{-}cracking}$ = 6.10 kN m/m and M_{creep} = 3.06 kN m/m. Then, these results were compared to those of a finite element model corresponding to the detailed design of a 2D-section of the Saint-Beat Tunnel with a layer of 10 cm of FRSC and steel arches spaced at 1.5 m intervals. The corresponding geological model is mostly composed of limestone (RMR IV according to the AFTES classification) modeled with a Mohr-Coulomb constitutive law, the parameters of which are γ = 26 kN/m^3, c = 150 kPa, φ = 35°, E = 1 GPa and υ = 0.3. The medium depth of the tunnel is about 120 m and the deconfinement rate is estimated at 20 % (λ = 0.80). After de-homogenisation, the results in terms of designed bending moment in the sprayed concrete layer ranged between 1 and 7 kN m/m, with a mean value of 4 kN m/m. Consequently, the loading rates considered in the Asquapro research programme are consistent with what can be observed in a tunnel project in a surrounding limestone rock with a low value of deconfinement. More detailed analyses are on-going.

Table 4 Duration of the pre-cracking phase

RSC	Type of reinforcement	Pre-cracking duration for specimens		
		A	B	C
BF01	Polymeric fibres	7 min	4 min	4.5 min
BF02	Steel fibres	–	–	–
BF03	Polymeric fibres	7 days	75 days	–
BF04	Steel fibres	6 min	–	–
BF05[a]	Polymeric fibres	25 min	2 min	1.5 min
BF06	Amorphous metallic fibres	–	–	–
WWM	Welded wire mesh	–	–	40 days

– means: the whole test
[a]This is a prototype fibre, not on the market

4.2 Pre-cracking Phase

During this phase (described in paragraph 2, stage 3), slabs were positioned in the creep test device; the applied load was $F_{pre\text{-}cracking}$. The load was maintained constant until the deflection (measured from the moment $F_{pre\text{-}cracking}$ was applied) exceeded 2 mm. For different types of reinforcement, the duration of this period varied a lot. For the 3 tested slabs of each RSC, results were either: consistent for BF01, BF02, BF05 and BF06; or highly variable for BF03, BF04 and WWM (see Table 4).

Most of the tested specimens reinforced with steel fibres, amorphous metallic fibres or welded mesh still had a deflection inferior to 2 mm under a load level of 120 % of the average maximum elastic force of their concrete matrix.

On the contrary, deflections of specimens reinforced with the tested polymeric fibres BF01 and BF05 exceeded 2 mm within a few minutes.

The behaviour of the specimens reinforced with the BF03 polymeric fibres was intermediate and varied from one specimen to another.

4.3 Mid-term Deflections Under Constant Load

The 0 point of the deflection curves was set after an offset of three minutes. This choice is due to the dispersion of the deflections recorded during the first minutes. This dispersion corresponds to the resetting of the slabs in their frame (after transportation from the punching press to the creep test apparatus), to the fibres distribution or mesh position and to the gradual application of the load as the screw is lowered. The deflection values during the first minutes cannot be analysed in a reliable way because all these parameters have not been investigated yet and because the duration of the load application may vary from one creep device to another. More work is needed on that subject.

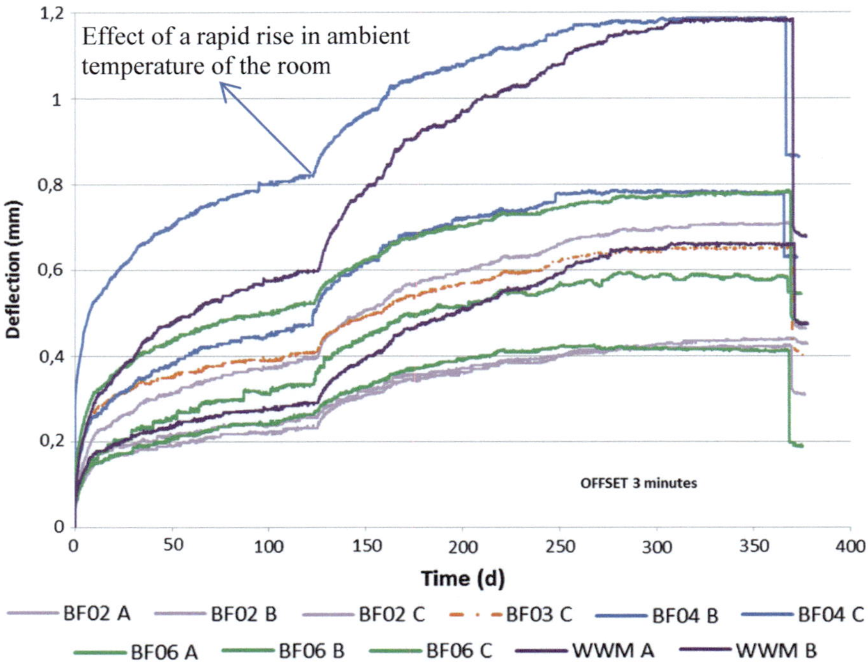

Fig. 2 Deflection versus time curves for slabs still in the pre-cracking phase

Figures 2 and 3 show the deflections of the slabs. They were separated into two sets: slabs still in the pre-cracking phase and slabs which load was reduced because their total deflection exceeded 2 mm (without the 3 min-offset). The first set of slabs (Fig. 2) were reinforced with metal fibres or welded wire mesh, apart from one slab reinforced with synthetic fibres; the second (Fig. 3) were reinforced with synthetic fibres, apart from two slabs—one reinforced with metal fibres and the other with welded wire mesh.

The shape of the curves, showing an inflection point around 120 days, is due to a rapid rise in ambient temperature, linked to repairs carried out on the heating system of the room. The acceleration of the deformations was observed during a rapid transition from 10° to 18 °C (see [2]).

Under their respective load (varying from 1 to 2 as said before), all slabs (except one) had relatively small deflection values (≤ 3 mm). The exception is a slab reinforced with synthetic fibres, which deflection was exponential and interrupted at around 20 mm when the loading plate came in contact with the loading screw (see Fig. 1).

One type of polymeric fibres provided reinforcement that is clearly superior to that provided by the two other polymeric fibres.

The behaviour of the metal fibres was more homogeneous but the variety of fibres tested is not sufficient to draw a general conclusion.

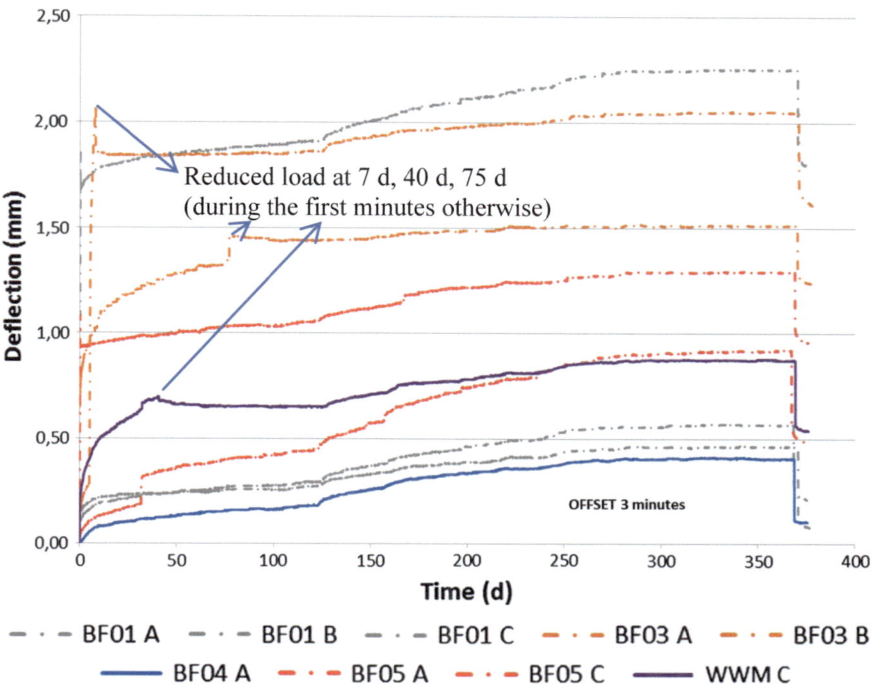

Fig. 3 Deflection versus time curves for slabs under creep load

The creep deformations of the two slabs reinforced with a welded wire mesh were of the same order as those reinforced with metal fibres; the third slab behaved in a similar manner to the slabs reinforced with synthetic fibres.

4.4 *Energy Absorption Capacity Before and After Creep Test*

The energy absorption capacity of nine specimens of each RSC was measured following the NF EN 14488-5: 3 specimens were tested 7 days after spraying, 3 others after 28 days, the last 3 were the slabs submitted to the creep tests (1 year of loading, 1 week unloaded).

BF06 fibres were not part of this comparison because they are specific fibres which are not primarily designed to provide ductility but to avoid first micro-cracks, strengthen structures and resist in aggressive ambient conditions.

The 3 main results from Table 5 are: (1) Even slabs with low pre-cracking duration and high creep deflection still had a good energy absorption capacity, which means that they still ensured the ductility of the material; (2) All RSC, except BF01, improved their energy absorption capacity, slightly or notably. (3) RSC

Table 5 Energy absorption capacity (J)

RSC	E 7d	E 28d	E post creep	Delta 7–28 d (%)	Delta pre-post creep (%)
BF01	859	877	750	2	−13
BF02	668	801	819	20	22
BF03	906	958	942	6	4
BF04	799	1045	838	31	5
BF05[a]	702	836	715	19	2
WWM	1029	1042	1478	1	44

[a]This is a prototype fibre, not on the market

BF02 and WWM showed the highest improvement of their E-values, these two reinforcements had the highest tensile strength (see Table 2), which confirms that higher reinforcement characteristics are required to improve ductility as the concrete matrix hardens.

The shapes of the curves are very different before and after creep (see Fig. 4a, b) as the slabs were already cracked in the second case. For example, BF05 does not

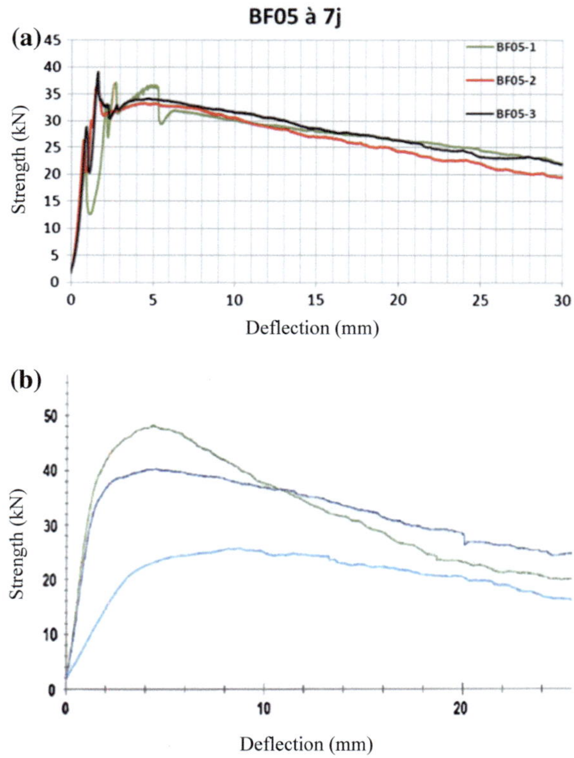

Fig. 4 **a** Strength-deflection curves of BF05 specimens before creep test. **b** Strength-deflection curves of BF05 specimens after creep test

show erratic pic values but a nearly perfect plastic behaviour. Note that specimen B of this FRSC had a deflection exceeding 20 mm, stopped when the loading plate came in contact with the loading screw.

4.5 Contribution of Reinforcements Prior to the Softening Phase

Figure 5 gives an example of the behaviour of a "good" FRSC submitted to a punching-bending load (EN 14488-5 test). Before the softening phase, 2 values are noteworthy: $F_{el\text{-}max}$, corresponding to the first drop in load and F_{max}, corresponding to the maximum value.

$F_{el\text{-}max}$ values were highly variable from one specimen to the other, even for a single RSC.

F_{max} values were much more consistent (see Table 6). They were measured at 7 days for all the specimens, reinforced or not, then at 28 days and after the creep tests but for the reinforced specimens only. Therefore, for each FRSC, results were analysed as a coefficient $C_{max} = F_{max(FRSC)}/F_{max(BB)}$ at 7 days, and then as a percentage of evolution with time.

All fibres had a positive impact on the maximum strength value; their impact was much lower than for the tested welded wire mesh. It reached up to 67 % for BF06 whereas it was only 14 % for BF05. This coefficient is useful when evaluating the contribution of fibres to the reinforcement of sprayed concrete.

Values were notably different from one fibre to another but they are not linked to the material of the fibre, either metallic or polymeric.

Evolution of F_{max} over time was positive from 7 to 28 days for all the tested RSC. F_{max} went on increasing for BF01, BF02, BF03 and WWM. Its evolution

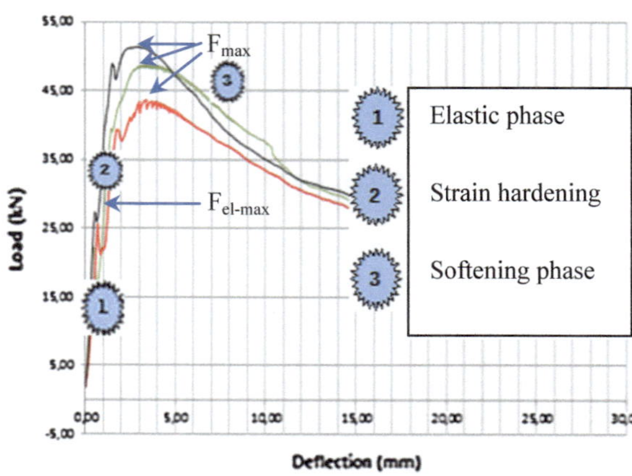

Fig. 5 Successive stages during loading

Table 6 Contribution of reinforcements prior to the softening phase

FRSC	F_{max} (7 d)	F_{max} (28d)	F_{max} (post creep)	C_{max} (7d)	Evolution F_{max} 7–28d (%)	Evolution F_{max} 28d-post creep (%)
BF01	49.89	54.67	56.33	1.51	10	3
BF02	44.93	54.00	60.67	1.36	20	12
BF03	47.62	48.33	54.00	1.44	1	12
BF04	47.96	67.00	58.00	1.45	40	−13
BF05	37.54	48.33	38.00	1.14	29	−21
BF06	55.22	58.00	54.00	1.67	5	−7
WWM	74.90	83.00	94.00	2.27	11	13

became negative for BF04, BF05 and BF06 but F_{max} values remain of the same order of magnitude (for BF05 & BF06) or higher than the 7d-value (for BF04).

Special attention must also be paid to the corresponding deflection values. Values obtained after creep are lower for metallic fibres than for polymeric fibres. A more in-depth analysis will be conducted at a later stage.

5 Conclusion, Perspectives

This experimental campaign on creep and energy absorption capacity is the first one dedicated to reinforced sprayed concrete. A new test method taking into account the specific support conditions of sprayed concrete for underground support has been designed and successfully experimented. More work is needed to assess the repeatability and reproducibility of this method.

During the experiment, six fibres and one welded wire mesh were tested. Three fibres were polymeric, three were metallic. Of course, not all the existing fibres on the market were covered and many parameters influence the results but conditions were selected to be as representative as possible of the current practise, at least in France. Still, results must not be extrapolated too quickly.

Under the level of load used during the pre-cracking phase (around 40 kN), the duration before the deflection exceeded 2 mm was of a few minutes for two out of the three polymeric fibres, whereas it was over one year for the three metallic fibres and the wire welded mesh. Results of the third polymeric fibre were intermediate (7 days; 75 days; over 1 year).

Under the load level of the creep tests (around 20 kN), no failure nor deflection above 3 mm (except for one sample) was observed.

All but one of the tested RSC maintained or improved their energy absorption capacity, even those with low pre-cracking durations or the highest creep deflection.

The contribution of reinforcements prior to the softening phase was quantitatively evaluated by calculating the ratio between F_{max} of each RSC and F_{max} of the non-reinforcement sprayed concrete with the same matrix. Values were notably different from one fibre to another (from +14 to +67 %) but they were not linked to the material of the fibre, either metallic or polymeric.

Acknowledgments The authors gratefully acknowledged the technical and financial supports from all the partners of this experimental programme: ASQUAPRO, ARCELORMITTAL, BASF DCCF, BEKAERT, BMS, CETU, ELASTOPLASTIC CONCRETE, NOUVETRA, SAINT GOBAIN SEVA, SIGMA BETON, SIKA, SNCF Réseau.

References

1. Larive, C., Rogat, D., Chamoley, D., Welby, N., Regnard, A.: Creep behaviour of fibre reinforced sprayed concrete. In: Proceedings of the World Tunnel Conference WTC 2015, May 22–28, Dubrovnik, Croatia
2. Larive, C., Rogat, D., Chamoley, D., Regnard, A., Pannetier, T., Thuaud, C.: Influence of fibres on the creep behaviour of reinforced sprayed concrete. In: Proceedings of the World Tunnel Conference WTC 2016 Congress, April 22–28, San Francisco, United States (submitted)

Effect of Beam Width on the Creep Behaviour of Cracked Fibre Reinforced Concrete

Raúl L. Zerbino, Graciela M. Giaccio, Diego H. Monetti and María C. Torrijos

Abstract Many efforts have been made to develop a creep testing procedure for fibre concrete in cracked state. Among several proposals the use of a bending test seems a promissory alternative. The "creep testing procedure" includes a pre-cracking process up to some established crack width, the creep test itself applying permanent loads and finally, a bending test to evaluate the residual strength properties after creep. This paper compares the results of creep tests performed on beams of different widths (50, 100 and 150 mm). The use of thin fibre reinforced concrete specimens should be of interest for applications such as the reparation of concrete structures or the protection of elements exposed to extreme actions. A conventional fibre concrete incorporating 40 kg/m^3 of hooked-end steel fibres with a 51.3 MPa compressive strength, class 4a, was used. The specimens were pre-cracked up to 0.5 mm and the sustained bending stress was equal to 60 % of the stress f_{R1} of each prism. The results show that the specimen width has a minor effect on the results of creep tests in cracked state but the variability increases when the width decreases.

Keywords Fibre reinforced concrete · Creep behaviour · Prism width · Steel fibres

1 Introduction

Creep characterization of cracked Fibre Reinforced Concrete (FRC) and the conditions required for a stable residual response represent a key point since serviceability of the material will depend on the stability of the cracks and the capacity to transfer sustained stresses.

R.L. Zerbino (✉) · G.M. Giaccio · M.C. Torrijos
LEMIT-CIC, Civil Engineering Department, UNLP, La Plata, Argentina
e-mail: zerbino@ing.unlp.edu.ar

D.H. Monetti
Civil Engineering Department, UNLP, La Plata, Argentina

P. Serna et al. (eds.), *Creep Behaviour in Cracked Sections of Fibre Reinforced Concrete*, RILEM Bookseries 14,
DOI 10.1007/978-94-024-1001-3_14

The *fib* Model Code 2010 [1] includes specific sections for FRC design. In many applications, and due to their benefits in crack control, fibres have been incorporated in concrete in order to improve the service life of the structures; thus, it must be emphasized that the material in service acts in a cracked state.

Many efforts have been made recently to develop a creep testing procedure for fibre concrete in cracked state and more than 15 research groups around the world are working on the subject at the moment. Among several proposals, the use of a bending test seems as a promissory alternative.

A creep bending test procedure has been proposed by Arango et al. [2]. The "creep testing procedure" includes a pre-cracking process up to some established crack width using the loading conditions indicated in the EN 14651 Standard [3], the creep test itself applying permanent loads with four-point loading conditions and finally, a bending test to evaluate the residual strength properties after creep. Using this method, Arango et al. [4] found low coefficients of creep at low loads and, clear effects of the type of fibre, the load level and concrete strength on the creep behaviour. More recently the method was applied to evaluate the creep and the residual properties in concretes reinforced with synthetic macrofibres exposed to different environmental conditions [5].

Using a similar methodology Zerbino and Barragán [6] studied the creep behaviour of steel FRC with initial crack openings of different sizes. In another paper, a clear effect of the initial crack size on the long term load capacity of a macro-synthetic FRC was observed [7]. A recent paper analysed some variables of the creep setup; no significant differences between creep tests performed with three or four-point loading configurations were found. Besides, the use of loading-unloading cycles does not contribute to the reduction of testing time [8].

This paper presents a study of the effect of the beam width (from 50 to 150 mm) on the creep behaviour in cracked state of FRC. The use of thin specimens should be of interest as they can be more representative of some applications such as the use of FRC for reparation of concrete structures or protection of elements exposed to extreme actions.

2 Experimental Program

A conventional fibre reinforced concrete incorporating 40 kg/m^3 of hooked-end steel fibres was used. A total of 18 prisms of 150 mm height and 600 mm length and different widths were cast; 6 of 50 mm, 6 of 100 mm and 6 of 150 mm width (the standard size). The specimens were turned on 90° for testing as it is usual. Standard 100 × 200 mm cylinders were also cast to evaluate the compressive strength. All specimens were cured in a moist room and after that they were placed two weeks in a laboratory environment to minimize the variations in strength during long term tests.

To characterize FRC bending behaviour one beam of each width was tested as indicated in the EN 14651 Standard [3] (three-point bending, 500 mm span and

25 mm notch). The crack mouth opening displacement (CMOD) was used as the control signal of a closed-loop servo-hydraulic system, through a clip gage placed 1 mm from the bottom of the beam. The first-peak strength (f_L) and the residual flexural strengths at 0.5, 1.5, 2.5 and 3.5 mm of CMOD, f_{R1}, f_{R2}, f_{R3} and f_{R4}, respectively, were calculated.

Four prisms of each width (named b–a to b–d, were "b" indicates the beam width in mm) were pre-cracked up to 0.5 mm, and then were used for creep tests. In these experiments, four-point loading configuration was adopted in order to improve the stability of the test; two beams were loaded together in each frame as shown in Fig. 1. For each pair of beams, the sustained load was calculated in order to apply a bending stress near to 60 % of the stress achieved at the end of the pre-cracking process, which corresponds to f_{R1} of each specimen.

To evaluate the CMOD during the creep test a mechanical dial gage was placed at the level of the tensile side of the beams (see detail in Fig. 1). The creep tests were conducted in an acclimatized room at 23 ± 2 °C.

The last prisms of each group were also pre-cracked up to 0.5 mm and remain unloaded with the purpose of comparing their mechanical properties with the residual strength properties of the specimens unloaded at the end of the creep tests. At the end of creep tests the content of fibres in the cracked sections were evaluated.

Fig. 1 Creep tests

3 Test Results and Analysis

3.1 Characterization Tests and Pre-cracking Process

Figure 2 shows the Stress–CMOD curves of the prisms -a to –d of each group during the initial cracking process (up to a CMOD of 0.5 mm) and the curves obtained during the bending tests performed in accordance to EN 14651 standard.

Table 1 presents the parameters obtained from the characterization tests (prisms –e) performed in accordance to the EN 14651 standard and the results of the pre-cracking process of the beams selected for the creep tests (prisms –a to –d). For each group of prisms the mean values and the COV are indicated, as well as those obtained from all specimens. Finally, the relative values of f_L, f_{max}, and f_{R1} of each width and the mean results of all specimens are calculated. It can be observed that there is a low variability in the measured stresses among the different groups of prisms width, and that the COV are as expected, lower than 10 % in the case of f_L and around 20 % for the residual stresses.

3.2 Creep Tests

Figure 3 shows the deferred crack opening under sustained loads versus time curves, for each specimen. Table 2 summarizes some characteristics of the creep tests including the instantaneous crack mouth opening after loading ($CMOD_i$), the applied sustained stress (f) and the relative value of the applied stress referred to f_{R1}, and two Crack Opening Rates ($COR_{60–30}$ and $COR_{90–60}$) calculated as chords considering the differed crack openings during the lapse of loading between 30 to 60 and 60 to 90 days respectively. After this period the specimens were unloaded and bending tests were performed to evaluate the residual strength properties after creep; Table 2 also shows the density of fibers on the fracture surfaces.

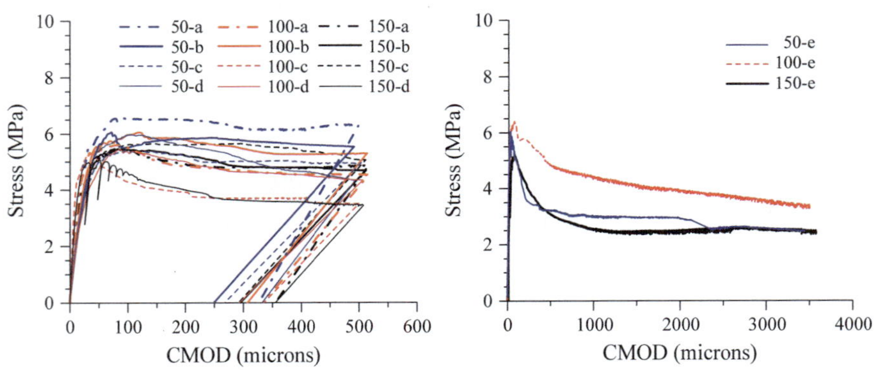

Fig. 2 *Left* pre-cracking process for creep tests (beams -a to -d); *right* EN 14651 tests (beams –e)

Table 1 Results of pre-cracking process and characterization tests (EN 14651)

Specimen	Width (mm)	h_{sp} (mm)	f_L (MPa)	f_{max} (MPa)	f_{R1} (MPa)	f_{R2} (MPa)	f_{R3} (MPa)	f_{R4} (MPa)
50-a	55	120	6.0	6.6	6.4			
50-b	59	120	5.6	6.1	5.5			
50-c	54	124	5.0	5.4	4.9			
50-d	60	121	5.4	6.0	4.3			
50-e	51	125	6.1	6.1	3.2	3.0	2.6	2.5
50-f	52	120	5.3	5.9	2.5			
50-mean	55	122	5.6	6.0	4.5			
50-COV	7	2	7	6	32			
100-a	108	120	5.5	5.6	4.5			
100-b	107	123	5.5	6.1	5.3			
100-c	104	122	5.3	5.3	3.7			
100-d	104	122	4.8	5.4	4.3			
100-e	98	124	6.1	6.4	4.8	4.1	3.8	3.4
100-f	100	118	6.0	6.0	3.2			
100-mean	104	122	5.5	5.8	4.3			
100-COV	4	2	9	7	18			
150-a	145	124	5.2	5.5	4.9			
150-b	148	125	5.2	5.5	4.7			
150-c	148	126	5.1	5.7	5.1			
150-d	147	126	5.2	5.2	3.4			
150-e	149	121	5.0	5.2	3.1	2.5	2.5	2.5
150-f	143	125	5.3	5.7	5.4			
150-mean	147	125	5.2	5.5	4.4			
150-COV	2	2	2	4	21			
Mean			5.4	5.8	4.4	3.2	2.9	2.8
COV			7	7	23	26	25	19
50-mean/mean			1.03	1.04	1.02			
100-mean/mean			1.02	1.01	0.98			
150-mean/mean			0.95	0.95	1.00			

It can be seen that most specimens show very similar creep behavior, with the exception of one atypical response in the case of beam 50-d. In the case of the curves of beams 100-b and 150-c, which have low f/f_{R1} ratios, the deformations are lower, while beam 150-d has the greatest deformations; however no significant differences in the crack opening rate related to the beam width are found. The variability of the response decreases as the beam width increases.

The evolution of crack opening along the time tends to be more stable after the first weeks under loading, showing homogenous increases after the first month; consequently after this age the creep response can be characterized by the concept of crack opening rate.

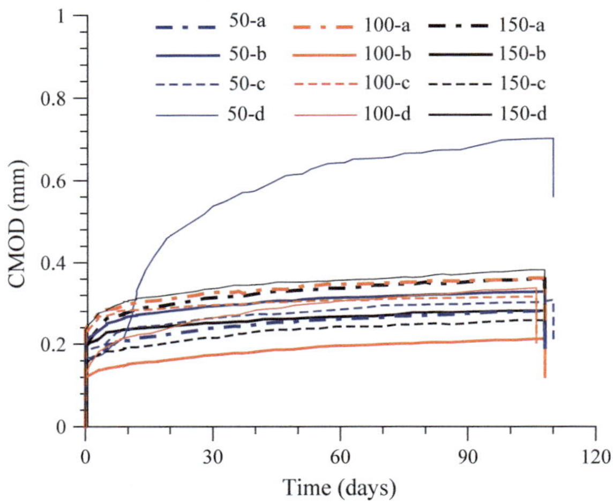

Fig. 3 Evolution of crack opening during creep tests

Table 2 Creep tests

Beam	$CMOD_i$ (mm)	f (MPa)	f/f_{R1} (%)	COR_{60-30} (mm/year)	COR_{90-60} (mm/year)	Fiber density (fibers/cm^2)
50-a	0.15	3.7	59	0.27	0.15	0.80
50-b	0.18	3.6	66	0.25	0.12	0.68
50-c	0.20	2.8	57	0.27	0.16	0.42
50-d	0.17	2.6	61	1.25	0.42	0.40
100-a	0.24	3.0	66	0.24	0.10	0.55
100-b	0.12	3.0	56	0.27	0.14	0.48
100-c	0.22	2.4	64	0.14	0.08	0.48
100-d	0.12	2.4	54	0.53	0.29	0.49
150-a	0.20	2.9	59	0.27	0.20	0.44
150-b	0.19	2.8	61	0.18	0.14	0.46
150-c	0.15	2.5	50	0.33	0.13	0.47
150-d	0.22	2.6	74	0.24	0.21	0.43

Figure 4 plots the $CMOD_i$ after loading as a function of the applied relative stress; it can be seen the consistency of the results between beams of different width. Figure 4 also shows the results of crack opening rate between 30 and 60 days, where, with the exception of the beam 50-d previously mentioned, there is a similar response with values near to 0.3 mm/year. In both cases the variability decreases in the standard beams of 150 mm width.

Finally, Fig. 5 plots the results of the crack opening rate during the 3rd month under loading (COR_{90-60}) versus those corresponding to the 2nd month (COR_{60-30}), it can be seen a decrease in the crack rate as expected, verifying that a very good correlation exists between those values with no differences regarding the width of the studied beams.

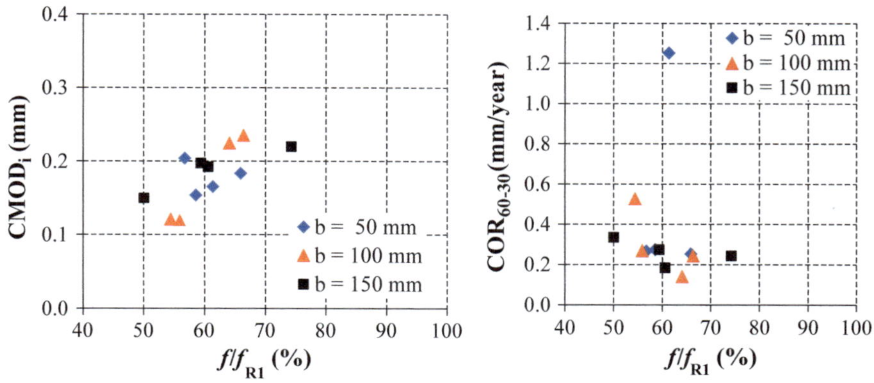

Fig. 4 Instantaneous crack opening after loading (*left*) and crack opening rate between 30 and 60 days (*right*) as a function of the relative applied stress

Fig. 5 Relationship between crack opening rates calculated for different times under loading

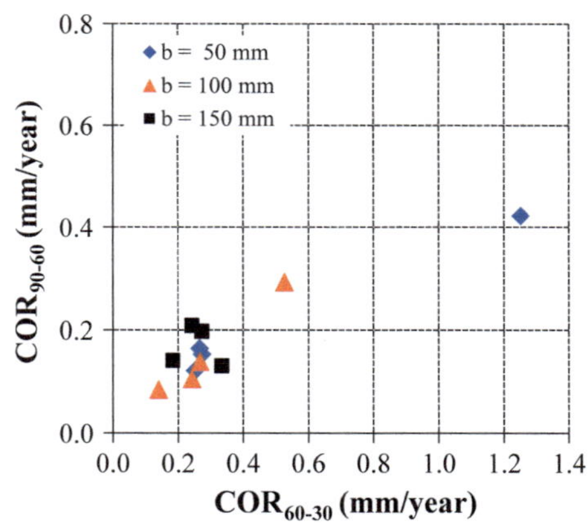

4 Effect of Long Term Loads on FRC Residual Strength Properties

After unloading bending tests were performed on prisms –a to –d and also on prisms –f (that remain unloaded) with the aim of evaluating the effect of long term loadings on the FRC residual strengths. Figure 6 shows the relationship between the residual stress f_{R3} and the residual stress f_{R1} obtained during the pre-cracking process. It can be seen a good correlation between both values and that long term loads did not produce any detrimental effect on the residual strength capacity; in addition, no differences between beams of different widths were found.

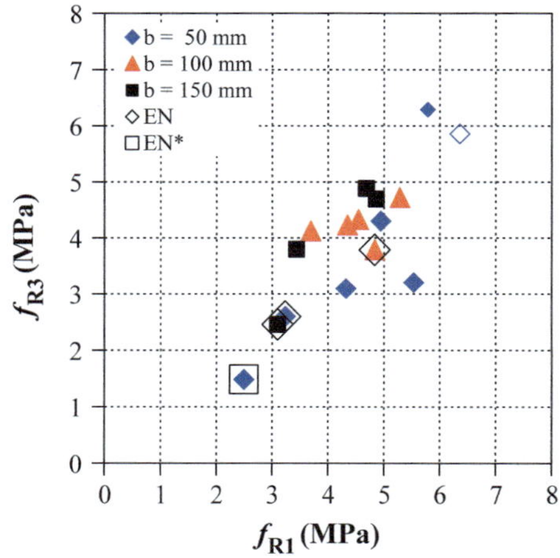

Fig. 6 Relationship between the residual stress f_{R3} measured during rapid bending tests after unloading and the residual stress f_{R1} obtained during the pre-cracking process (◇ initial characterization tests, prisms –e; specimen not submitted to creep tests, prisms –f)

5 Effect of the Content of Fibres in the Cracked Section

After bending tests, the content of fibres in the fractured surfaces of each specimen was counted. In this way, it can be discussed if the beam width can affect the creep results due to variations in the fibre density or fibre orientations.

Figure 7 shows the variation in the mechanical properties measured in bending with the fibre density measured at the fracture surfaces. As it can be seen the density of fibres does not affect the first peak load but it has a significant influence on the residual stress before (f_{R1}, Fig. 7 left) and after creep tests (f_{R3}, Fig. 7 right); in both cases the same tendency can be seen for prisms of different widths but it must be noted that the greatest variations in fibre density correspond to the thinner beams. Thus, the greater variability in the results from the 50-mm width beams is clearly justified. Analysing Table 2, it can be seen that the COV of the fibre density is only 4 and 7 % in the case of 150-mm and 100-mm beams and it increases up to 34 % in the smallest beams.

Regarding the effects of fibre density at the fracture surfaces on the results of COR, it can be seen in Fig. 8 that the highest values of crack opening rate are associated to low fibres contents; nevertheless it must be remembered that the applied sustained loads were related to the residual capacity measured during pre-cracking in each pair of companion beams (near 60 % f_{R1}). Regardless these observations and for these loading conditions, Fig. 8 shows that the variation in crack opening rate is not so much affected by the beam width and the fibre density and then, the results of a same concrete obtained from beams of different width are comparable.

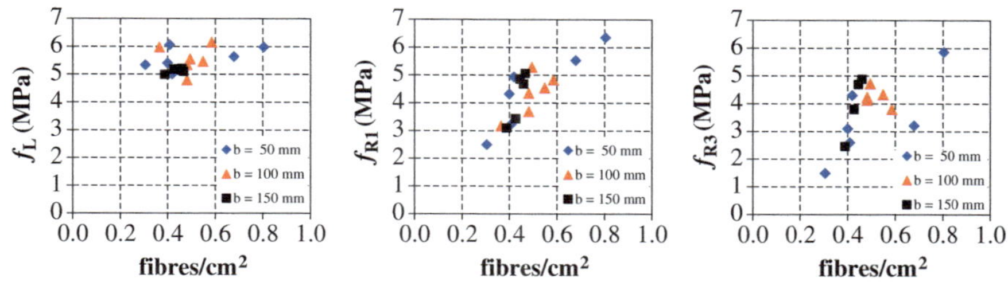

Fig. 7 Variation of mechanical properties measured in bending with the fibre density. *Left* and *centre* pre-cracking tests; *right* bending tests after creep loading

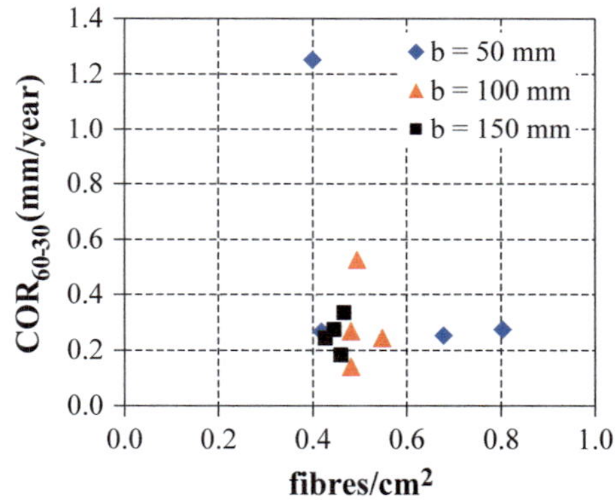

Fig. 8 Relationship between fibre density and crack opening rate

6 Conclusions

This paper compares the results of creep tests performed on beams of different widths (50, 100 and 150 mm); the following conclusions can be drawn:

- Stable creep tests were observed when sustained stresses near 60 % f_{R1} were applied on a “class 4a” steel FRC.
- The results show that the beam width has a minor influence in the creep behaviour of FRC in cracked state, thus the proposed method can also be applied to evaluate thin specimens.
- A greater variability was found in the fibre density on the cracked surfaces as the beam width decreases; as expected the residual strength capacity of the beams was directly related to the fibre density.
- The variability in creep behaviour appears to decrease as the beam width increases. Therefore an increase in the number of specimens is recommended in the case of thin beams.
- The concept of crack opening rate appears as a useful tool to evaluate the creep behaviour in cracked FRC.

Acknowledgments Funding from two projects, CONICET PIP 112-201101-00765 "Development and characterization of fibre reinforced concretes for structural applications", and UNLP 11/I188 "Fibre reinforced concretes and their contribution to the sustainable development" is greatly appreciated. The authors specially thank the collaboration of Jorge Anile, Teobaldo Gaitán, Anabela Gerez and Pablo Bossio in the support of experimental works.

References

1. Fédération Internationale du Béton. *fib* Model Code for Concrete Structures 2010. Berlin, Germany (2013)
2. Arango, S., Serna, P., Martí Vargas, J.R., García Taengua, E.: A Test method to characterize flexural creep behaviour of pre-cracked FRC specimens. *Exp. Mech.* **52**(8), 1067–1078 (2012)
3. European Committee for Standardization: EN14651:2005 Test Method for Metallic Fibered Concrete—Measuring the Flexural Tensile Strength (Limit of Proportionality (LOP), Residual). Brussels, Belgium (2005)
4. Arango S., Taengua, E.G., Martí Vargas, J.R., Serna Ros, P.: A comprehensive study on the effect of fibers and loading on flexural creep of SFRC. RILEM PRO88, Proceedings of the 8th RILEM Int. Symp on Fibre Reinforced Concrete, Guimaraes (2012)
5. Serna Ros, P., Martí Vargas, J.R., Bossio M.E., Zerbino, R.: Creep and residual properties of cracked macro-synthetic fibre reinforced concretes. Mag. Concr. Res. (2015). doi:http://dx.doi.org/10.1680/macr.15.00111
6. Zerbino, R., Barragán, B.: Long-term behavior of cracked steel fiber-reinforced concrete beams under sustained loading. ACI Mater. J. **109**(2), 215–224 (2012)
7. Monetti, D.H., Giaccio, G., Zerbino, R.: Fluencia en estado fisurado en hormigón reforzado con macrofibras sintéticas. In: VI Congreso Internacional, 20ª Reunión Técnica AATH, Concordia, Argentina, pp. 135–142 (in Spanish) (2014)
8. Zerbino, R., Monetti, D.H., Giaccio, G.: Creep behaviour of cracked steel and macro-synthetic fibre reinforced concrete. Mater. Struct. (2015). doi:10.1617/s11527-015-0727-y

Macro-Synthetic Fibre Reinforced Concrete: Creep and Creep Mechanisms

A.J. Babafemi and W.P. Boshoff

Abstract The creep of cracked macro-synthetic fibre reinforced concrete has been investigated in uniaxial tension at five stress levels (30, 40, 50, 60 and 70 %) of the residual uniaxial tensile strength under controlled temperature and relative humidity of 23 ± 1 °C and 65 ± 5 % respectively. Specimens were prepared and cracked before they were subjected to the various stress levels. One mix design has been used throughout the investigation and the fibre was used at a volume of 1 %. Test results of the investigation has revealed that while the creep response is a function of the applied stress level, significant creep has been recorded after 8 months even at lower stress levels. The mechanism responsible for the time-dependent crack opening of macro-synthetic fibre reinforced concrete was investigated through time-dependent fibre pull-out and fibre creep tests. The result of this investigation has shown that macro-synthetic fibre creep and the time-dependent fibre pull-out are mechanisms responsible for the creep of cracked macro-synthetic fibre reinforced concrete.

Keywords Fibre reinforced concrete · Creep · Creep mechanism · Macro-synthetic fibre · Stress level · Time-dependent deformation

1 Introduction

Usually, structures are subjected to a combination of permanent and imposed loads throughout its service life. The time-dependent deformation of any part of the structure due to loading is regarded as creep. All concrete structures undergo creep as soon as they are subjected to sustained loadings (change in strain with time under sustained stress). If not subjected to sustained loadings, they still undergo some form of deformation which is known as shrinkage [1]. Many studies have been undertaken in the past to understand the creep of plain concrete, the mechanisms

A.J. Babafemi · W.P. Boshoff (✉)
Department of Civil Engineering, Stellenbosch University, Stellenbosch, South Africa
e-mail: bboshoff@sun.ac.za

P. Serna et al. (eds.), *Creep Behaviour in Cracked Sections of Fibre Reinforced Concrete*, RILEM Bookseries 14,
DOI 10.1007/978-94-024-1001-3_15

responsible for its time-dependent deformation and the factors influencing creep behaviour of concrete. Results of such studies have attributed the time-dependent deformation of concrete to the volumetric change in the cement paste content under applied load [2–4]. The rate of the volumetric change in the cement paste is a function of a host of factors which include the water-cement ratio, environmental conditions (temperature and humidity), applied stress level and volume and properties of aggregates used [1]. Based on the type of loading condition (compression and tension), some authors have shown that for mass concrete, uniaxial tensile creep is somewhat 20–30 % higher than specimens tested in uniaxial compression under the same stress level [1]. The creep of unreinforced concrete can lead to cracking, excessive deflection and other serviceability and durability issues.

On the other hand, when reinforcement such as fibres, which are now being increasingly used for reinforcing structures such as ground slabs, industrial floors, bridge deck, pipes and tunnel lining [5, 6] are introduced into plain concrete, the study of its creep behaviour takes a whole new dimension. Fibres are generally known to bridge cracks when they are formed in concrete, thereby leading to significant residual tensile strength of concrete in the post-crack state. Due to the action of the fibres bridging the cracked planes, energy absorption capacity is also significantly enhanced. With fibres having diameter less than 0.3 mm, improved durability properties of concrete have been reported due to reduction in crack width and permeability of harmful substances [7].

To be able to understand the time-dependent deformation of concrete reinforced with fibres otherwise called fibre reinforced concrete (FRC), tests would have to be performed in the cracked state as the fibres are only functional as crack-bridging material in the cracked state. Though a number of guidelines exist on the reasonable use of fibres in concrete, there is yet to be a guide on the creep behaviour of FRC in the cracked state. This has necessitated the recent rise in research activity on the creep of cracked FRC. Some of the leading works in this respect are those by [8–17]. Whereas most of these works were carried out using steel FRC, only a handful have considered the creep of cracked macro-synthetic FRC. Also, studies on the uniaxial tensile creep of cracked macro-synthetic FRC are still very limited. One of the first published work on the creep of cracked macro-synthetic FRC is that reported by Babafemi and Boshoff [8].

Macro-synthetic fibres (diameter $\geq$ 0.3 mm) are becoming increasingly preferred to steel fibres due to several reasons: easy to work in concrete, lightweight, resistance to corrosion and cheaper in cost. However, macro-synthetic fibre has an elastic modulus (3–7 GPa) far less than that of steel fibre (210 GPa) and responds in a viscoelastic manner (being a plastic fibre) under load. If this fibre is to be safely used for structural concrete, its long-term behaviour in the cracked form under sustained loadings must be understood. For cracked specimens of macro-synthetic FRC tested under sustained loadings, possible failure modes would be fibre pull-out and fibre rupture. An understanding of the mechanism responsible for the failure is also germane to a study of the time-dependent behaviour of cracked macro-synthetic FRC. Therefore, this study examines the creep response of cracked macro-synthetic FRC and the mechanisms responsible for the creep behaviour.

2 Materials and Methods

2.1 Fibre and Concrete

The macro-synthetic fibre used for this investigation is polypropylene having the properties shown in Table 1. The fibre is slightly crimped and designed to have a star cross section area so as to enhance frictional bond with the cementing matrix over its full length.

For the purpose of this study, a concrete mix with a high slump was developed to make allowance for the reduction in slump that the addition of the fibre will create. At least 90 % of the coarse aggregate (crushed Greywacke stone) used passes through the 6 mm sieve size and retained on a 4.75 mm sieve while the fine aggregate was a natural sand (locally known as Malmesbury sand) passing through 4.75 mm sieve with a fineness modulus (FM) of 2.34. CEM I 52.5 was used as the binder. Fibre content for all mixes was kept constant at a volume of 1 %. Details of the concrete mix used is presented in Table 2. When fibres are added all at once to the concrete during mixing, they have the tendency for balling, hence the fibres were carefully added to the concrete mix until a visual uniform distribution of the fibres had been observed. The consistence of the mixes (before and after the addition of fibres) were measured using a standard slump test [18]. The slump before the addition of fibres was 160 mm and reduced to 100 mm after the addition of the fibres.

Table 1 Properties of macro-synthetic fibre [19]

Specific gravity	0.88–0.92
Modulus of elasticity	4.3 GPa
Colour	Translucent
Elongation at yield	15–25 %
Tensile strength	400 MPa
Melting point	150–170 °C
Length (l_f)	40 mm
Nominal diameter (d_f)	0.8 mm
Effective diameter (d_{eff})	0.794 mm
Aspect ratio (l_f/d_f)	50

Table 2 Concrete mix design

Material type	kg/m^3
Cement (CEM I 52.5)	395
Stone (Greywacke = 6 mm)	800
Sand (Malmesbury)	990
Water	190
Polypropylene fibre (1 % by volume)	9.1
Superplasticizer (0.2 % by weight of binder)	0.79

2.2 Mould and Specimen Preparation

The test specimens used for the investigation of the uniaxial tensile creep of cracked macro-synthetic FRC were prepared from modified steel prism moulds measuring $100 \times 100 \times 500\ mm^3$ as shown in Fig. 1.

Specially designed steel hooks were worked into the moulds for the purpose of supporting the specimens in the testing machine. Adequate care was taken to ensure the alignment of the central rods of the anchors positioned at each end of the mould to avoid eccentricity as much as possible. After the horizontal and vertical alignments of the central rods of the anchors were achieved, they were then supported in position with a crossbar using a binding wire already attached to the central rod of each anchor.

After the preparation of the moulds, specimens were cast and compacted using a vibrating table. The cast specimens were demoulded after 21 ± 1 h and cured by complete immersion in water at a controlled temperature of 23 °C. Concrete cubes (100 mm) with and without fibres were also cast to determine the compressive strength of the concrete mixes. The compressive strength of the mixes, with and without fibres are 40.19 and 43.80 MPa respectively.

Fig. 1 Modified steel moulds showing anchors [8]

2.3 Uniaxial Tensile Strength Test

After demoulding the specimens on the 28-day of casting and curing, the specimens were then prepared for the uniaxial tensile test. The test was carried out to determine the residual uniaxial tensile strength of the specimens as the tensile creep stress levels were based on percentages of the average residual tensile strength. The specimens were notched at the centre on all four sides to a depth of 10 mm using a diamond blade with a width of 3 mm. This gave an effective cross-sectional area of $80 \times 80\ \text{mm}^2$ at the centre of each notched specimen. For the uniaxial tensile test, three specimens were used and an average residual tensile strength was determined. The test was performed in a Zwick Z250 Universal Testing Machine as shown in Fig. 2. To measure the crack mouth opening displacement (CMOD) during the test, two linear variable displacement transducers (LVDT) were mounted on aluminium frames (gauge length, 120 mm) attached to the specimen and readings were saved throughout each test. The test was performed at a crosshead displacement rate of 0.01 mm/s; adequate care was ensured that cracking did not occur before the test was started.

Fig. 2 Uniaxial tensile strength test of macro-synthetic FRC [8]

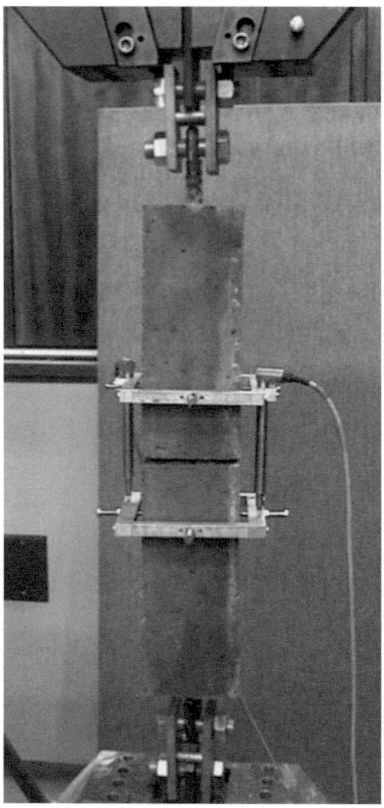

2.4 Uniaxial Tensile Creep Test

After the determination of the average uniaxial tensile strength, the creep test was then pursued. Specimens for the creep test had been cracked using the same procedure and crosshead displacement rate for the uniaxial tensile test. The specimens were cracked to an average crack width of 0.5 mm, unloaded and transferred to the creep frame where they were subjected to creep stresses of 30, 40, 50, 60 and 70 % of the average residual tensile strength. The creep frame was designed to accommodate two specimens in series as shown in Fig. 3.

Prior to subjecting the specimens to the creep loads using a lever arm system, the creep frames were calibrated using a 50 kN tensile load cell to determine the precise loads to be supported (including self-weight of lever arms and suspension cables) at each stress level. Table 3 shows the details of the various stress levels tested.

The result of the tensile creep measured in terms of CMOD for 8 months was measured using a 10 mm LVDT (two for each specimen) mounted in the same fashion as specimens for the uniaxial tensile strength test. It should be noted that unlike the uniaxial tensile strength test specimens where a gauge length of 120 mm was used, specimens tested for creep had a gauge length of 70 mm. This was due to

Fig. 3 Tensile creep frames showing specimens under sustained loadings [8]

Table 3 Applied stress levels and number of specimens tested

Load levels as % of the post-crack strength	30 %	40 %	50 %	60 %	70 %
Actual applied stress (MPa)	0.234	0.312	0.39	0.468	0.546
Equivalent creep load (N)	1498	1997	2496	2995	3494
Number of specimens	4	4	4	2	2

the smaller LVDTs used for the creep test. All tests were carried out in a controlled climate room where temperature and relative humidity were maintained for the test period at 23 ± 1 °C and 65 ± 5 % respectively.

The displacement reading obtained from the creep test will include material creep and drying shrinkage. The specimens were tested unsealed, therefore basic creep, drying creep, crack widening and drying shrinkage occurred at the same time. The drying shrinkage could be subtracted as it was measured separately while the material creep is insignificant compared to the crack widening.

2.5 *Drying Shrinkage Test*

The drying shrinkage was determined from two load free specimens of the same size and geometry as those used for the creep test. Two LVDTs extended to have a gauge length of 285 mm (more than half the specimen length) were attached to the specimens. The readings were measured electronically using a HBM Spider8 Electronic Measuring System. Since the gauge length was different from that of the creep specimens, shrinkage strain measured was multiplied by the gauge length of the creep specimens to determine the displacements caused by shrinkage in the creep specimens over a gauge length of 70 mm. A deduction of these strains from the measured CMOD gave the actual creep of the specimens under sustained loading. Figure 4 shows the shrinkage specimens during testing.

Fig. 4 Drying shrinkage test specimens [8]

2.6 Single Fibre Tests

To understand the mechanisms responsible for the creep of cracked macro-synthetic FRC subjected to sustained tensile loading, time-dependent fibre pull-out and fibre creep tests were performed. The time-dependent fibre pull-out test was pursued by preparing fresh concrete using the mix design in Table 2 without the inclusion of fibres. 100 mm cube specimens were then cast and single macro-synthetic fibres (already marked) were inserted by hand to a depth of 25 mm into the fresh concrete mix at the marked mid-point of the surface area of the cast concrete. The moulds were then gentle vibrated on a vibrating table to ensure closure of any voids created during the insertion of the fibres. Care was taken to ensure that the fibres remain vertical after the gentle vibration of the moulds. After demoulding and curing for 28 days, the specimens were prepared for testing. A hole of 12 mm diameter was drilled on the opposite side to where the fibre was embedded to a depth of about 30 mm. Threaded bars with 10 mm diameter to support the specimens were then fixed in the holes with epoxy glue and left in the controlled climate room to set for 24 h before testing commenced. Specimens were then positioned in a frame for testing as shown in Fig. 5. The free ends of the fibres were gripped with a hand drill chuck and predetermined sustained loads between 50 and 80 % of the average interfacial shear resistance of specimens tested at 25 mm embedment length were applied to the fibres individually to study the time-dependent pullout behaviour.

Measurement of the fibre pullout over time was captured optically with the aid of a microscope using a 3.1 Mega-pixel Leica EC3 camera. Before the application of the weights, a photo of the setup was taken to serve as the reference photo. The height of the mouth grip of the hand drill chuck, h_{ch}, 5 mm, was used as a scale reference (Fig. 5). With this scale, the reference fibre length (from reference photo) before pullout, L_r, was determined. Subsequently, photos were taken at regular

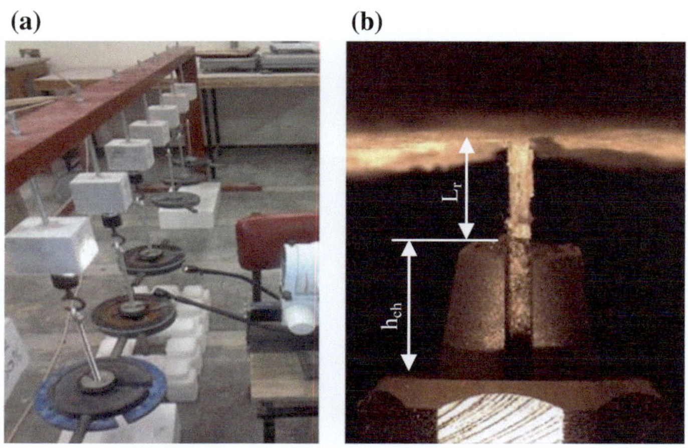

Fig. 5 Test setup for time-dependent fibre pull-out [8]

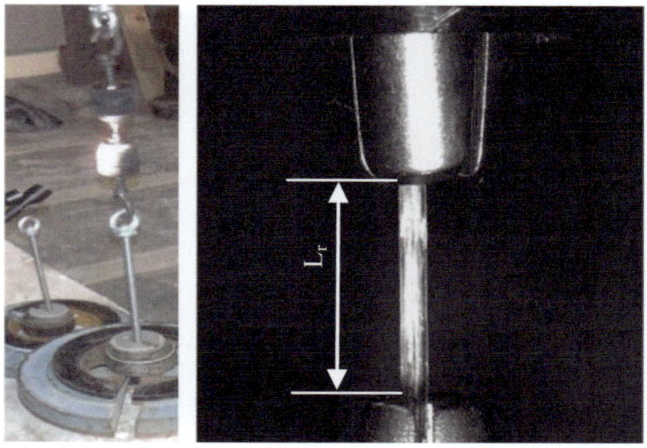

Fig. 6 Fibre creep test setup [8]

intervals of the fibre pullout from the matrix and the new pullout length was then subtracted from L_r to obtain the time-dependent fibre pullout displacement.

The fibre creep test was carried out to understand the creep mechanism of cracked macro-synthetic FRC. Since this type of fibre is well known to creep, its contribution to the overall creep of the specimens is sought. One fibre was used for this investigation. Hand drill chucks of the same type as those used for the time-dependent fibre pull-out test were employed to grip the fibre at its free ends. Damage to the gripped ends of the fibre by the chucks and slippage during test did not occur. Where slippage was noticed, specimens were discarded. Thereafter the specimen was supported by a frame as that in the time-dependent pull-out test and the lower chuck supported the sustained load as shown in Fig. 6.

The same procedure for measuring the time-dependent pullout test was also used to determine the time-dependent elongation of the fibre under sustained load of 143 MPa. This represents 30 % of the average tensile strength of the fibres tested in tension at a rate of 0.5 mm/s.

3 Experimental Results and Discussion

3.1 Uniaxial Tensile Strength Result

The result of the tensile test for the three specimens tested is presented in Fig. 7. The result shows the stress (σ) versus CMOD (w) relationship. The result shows an average ultimate tensile strength of 2.90 MPa while the average residual tensile strength at the point of full activation of the fibres is 0.78 MPa at a crack width of about 0.6 mm. Significant crack opening is observed as soon as the specimens cracked (brittle failure with loud noise) at the ultimate tensile strength. This significant crack opening before the fibres were engaged to control further opening is due to the lack of adequate stiffness from the testing machine. For such tests, a test

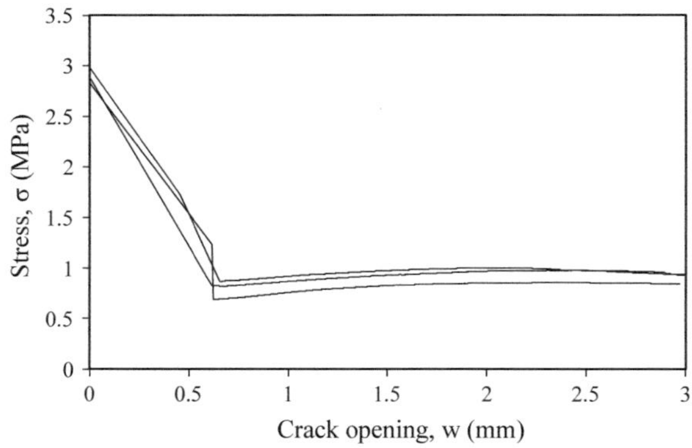

Fig. 7 Stress-crack opening of tensile test [8]

setup with high stiffness and a servo-controlled closed loop control is more appropriate and will give better results. Secondly, due to the plastic nature of macro-synthetic fibre, some level of stretch is expected before fully being engaged to control crack widening. However, for the purpose of this study, the average residual tensile strength is what is of interest. The creep loads were then based on the average residual strength of 0.78 MPa as depicted in Table 3.

3.2 Uniaxial Tensile Creep Result

After subjecting the specimens to the respective creep loads for 8 months, Fig. 8 shows the results of the average time-dependent crack opening at the various stress levels investigated. It should be noted that the effect of shrinkage has already been deducted from the overall creep measured as previously mentioned.

It is acknowledged that there is variability in the test results of specimens within each stress level due to the distribution and number of fibres, however, the goal here is to consider the average creep response based on the stress levels tested. Evidently, from the result presented in Fig. 8, the creep response is stress dependent. The higher the stress level, the higher the creep measured as crack opening. At higher stress levels (60 and 70 %), the creep was significantly pronounced leading to creep fracture after less than a day for one of the two specimens subjected to 70 % stress level. Similarly, one of the specimens tested at 60 % stress level was also unloaded from its creep frame after 15 days when the crack opening had reached the maximum measuring capacity of the LVDTs (10 mm). So, results for 60 and 70 % stress levels are not averages but the results of the single specimen left. At 40 % stress level, the result represented the time-dependent behaviour up to 6 months because the test was started much later from the others. However, the crack opening still lie between that of 30 and 50 % stress levels and expected would have followed the same trend up to 8 months.

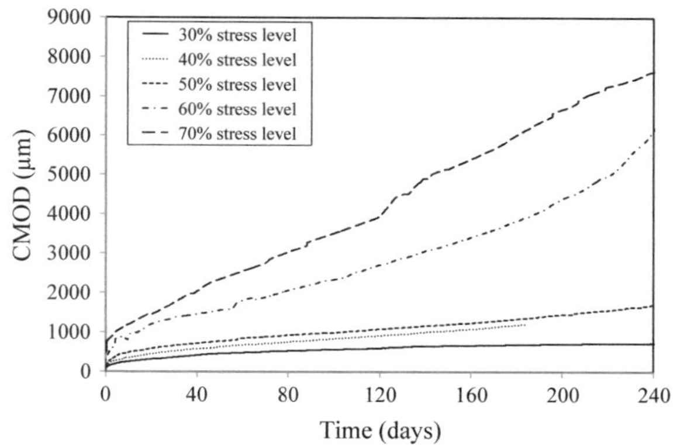

Fig. 8 Average tensile creep results at different stress levels [8]

It is noteworthy to observe that, the results presented in Fig. 8 indicate that even at the lowest stress level of 30 % (i.e. 0.23 MPa), a total crack opening of 1.2 mm is recorded, which is quite significant. This means that even at such low levels of loading, the crack width increase is significant during uni-axial sustained loading.

3.3 *Single Fibre Tests*

The results of the time-dependent fibre pull-out and fibre creep tests have shown that these phenomena are majorly responsible for the creep of cracked macro-synthetic FRC. Figure 9a, b shows the results of the time-dependent pull-out and fibre creep tests. For the pull-out test, all the fibres pulled out of the specimens at all load level without any fibre rupture. It is also evident from Fig. 9a that the fibre pull-out creep is also dependent on the stress level.

The result of the single fibre creep test (Fig. 9b) shows significant creep of 40 % strain in less than 4 days at a load of 30 % of the fibre's tensile capacity. Due to

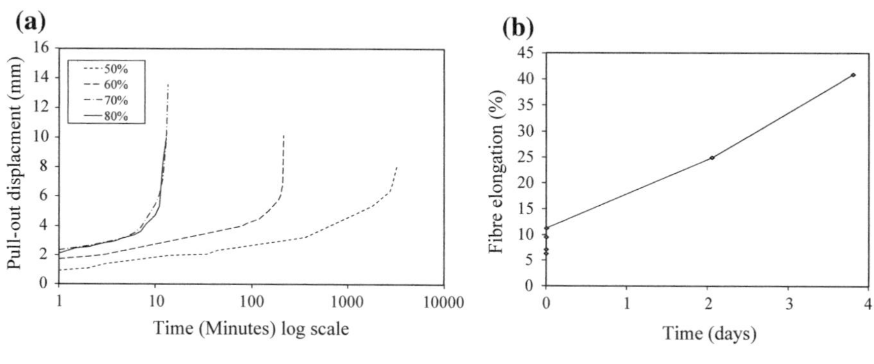

Fig. 9 Time-dependent fibre pull-out **a** and fibre creep **b** results [8]

Poisson's effect, this lengthening will result in the fibre's cross section reducing over time. The transfer of load between the fibre and the matrix when the matrix is cracked is dependent on the frictional bond and mechanical interlock between the fibre and the matrix. When the fibre contracts laterally, both the frictional bond and the mechanical bond will weaken and this will result in partial fibre pull-out. This process will continue until this debonding stabilises or the fibre pulls out completely which in turn leads to failure. Hence, it is evident that the creep of the fibre is a contributor to the crack widening of cracked macro-synthetic FRC under sustained loading which results in partial of full fibre pull-out.

4 Conclusions

- Cracked macro-synthetic fibre reinforced concrete (FRC) has shown significant tensile creep over 8 months even at a stress level as low as 30 % of the average residual tensile strength. Creep should therefore be taken into account in the use of macro-synthetic FRC.
- Creep fracture of specimens occurred at 60 and 70 % stress level; hence, such stress levels are not sustainable for cracked macro-synthetic FRC.
- Mechanisms responsible for the creep of cracked macro-synthetic FRC are the time-dependent fibre pull-out and fibre creep under sustained loadings.

References

1. Neville, A.: Properties of concrete. In: 5th & Final Edn. Pearson Education Limited, England (2011)
2. Kumar, G.A., Kumar, S.N.: Creep of Concrete. Int. J. Eng. Dev. Res. (IJEDR) **2**(4), 3800–3802 (2014)
3. Bazant, Z.P., Santosh, P.: Solidification theory for concrete creep. II: Verification and application. Journal of Engineering mechanics **115**(8), 1704–1725 (1989)
4. Bažant, Z.P., Hauggaard, A., Baweja, S., Ulm, F.: Microprestress-solidification theory for concrete creep. I: aging and drying effects. J. Eng. Mech. **123**(11), 1188–1194 (1997)
5. Buratti, N., Mazzotti, C., Savoia, M.: Post-cracking behaviour of steel and macro-synthetic fibre-reinforced concretes. Constr. Build. Mater. **25**(5), 2713–2722 (2011)
6. Oh, B.H., Park, D.G., Kim, J.C., Choi, Y.C.: Experimental and theoretical investigation on the postcracking inelastic behaviour of synthetic fibre reinforced concrete beams. Concr. Cement Res. **35**, 384–392 (2005)
7. Brandt, A.M.: Fibre reinforced cement-based (FRC) composites after over 40 years of development in building and civil engineering. Compos. Struct. **86**(1), 3–9 (2008)
8. Babafemi, A.J., Boshoff, W.P.: Tensile creep of macro-synthetic fibre reinforced concrete (MSFRC) under uni-axial tensile loading. Cem. Concr. Compos. **55**, 62–69 (2015)
9. Zerbino, R.L., Barragan, B.E.: Long-term behaviour of cracked steel fibre-reinforced concrete beams under sustained loading. ACI Mater. J. **109**(2), 215–224 (2012)
10. Bernard, E.S.: Creep of Cracked Fibre Reinforced Shotcrete Panels', in 'Shotcrete: More Engineering Developments. Taylor and Francis Group, London (2004)

11. MacKay, J., Trottier, J.F.: Post-Crack Creep Behavior of Steel and Synthetic FRC Under Flexural Loading', in 'Shotcrete: More Engineering Developments. Taylor and Francis Group, London (2004)
12. Kurt, S., Balaguru, P.: Post crack creep of polymeric fibre-reinforced concrete in flexure. Cem. Concr. Res. **30**(2), 183–190 (2000)
13. Bernard, E.S.: Influence of fiber type on creep deformation of cracked fiber-reinforced shotcrete panels. ACI Mater. J. **107**, 474–480 (2010)
14. García-Taengua, E., Arango, S., Martí-Vargas, J.R., Serna, P.: Flexural creep of steel fiber reinforced concrete in the cracked state. Constr. Build. Mater. **65**, 321–329 (2014)
15. Arango, S.E., Serna, P., Martí-Vargas, J.R., García-Taengua, E.: A test method to characterize flexural creep behaviour of pre-cracked FRC specimens. Exp. Mech. **52**, 1067–1078 (2012)
16. Zhao, G., di Prisco, M., Vandewalle, L.: Experimental investigation on uniaxial tensile creep behavior of cracked steel fiber reinforced concrete. Mater. Struct. **48**, 3173–3185 (2014)
17. Buratti, N., Mazzotti, C.: Experimental tests on the effect of temperature on the long-term behaviour of macro-synthetic fibre reinforced concretes. Constr. Build. Mater. **95**, 133–142 (2015)
18. BS EN 12350-2, 'Testing fresh concrete – Slump test', (BSI, London, 2000).
19. Rocstay. Polypropylene fibres for concrete and mortar reinforcement. Technical Data sheet. A publication of Fibsol (Fibre reinforcing solution), South Africa.

Part V
Influence of Applied Load on Creep Tests

Influence of Fibre Reinforcement on the Long-Term Behaviour of Cracked Concrete

Aitor Llano-Torre, Samuel Eduardo Arango, Emilio García-Taengua, José Rocío Martí-Vargas and Pedro Serna

Abstract The influence of fibre reinforcement on the long-term behaviour of cracked concrete is analysed in this work by means of a creep test. Nine concrete mixes were prepared (7 SFRC's and 2 conventional RCs) based on two basic mix designs. Concretes type I were conceived for structural pre-cast applications and concretes type II reproduce a general purpose. Fibre dosages and conventional reinforcements were varied to represent a wide spectrum of post-peak flexural responses. In all cases with fibre reinforcement steel fibres were used. Conventional RC specimens were reinforced with two steel rebars. In addition to the variables of mix design of concrete, there are two significant variables related to the creep test: the pre-crack opening level ($CMOD_{pn}$) and the stress level (I_c) sustained during the test. Creep tests were performed by applying a constant flexural load on notched pre-cracked specimens and controlling crack opening evolution. Some of the specimens developed a sudden increase of crack opening deformations during the creep test. Creep coefficients and Crack Opening Rates were calculated and analysed. Creep coefficients show significant dependence on the analysed variables. The results of this experimental campaign show that creep on SFRC specimens may be similar to a conventional RC.

Keywords Creep · Steel fibre reinforced concrete · SFRC · Steel fibres · Long-term · Crack opening

---

A. Llano-Torre (✉) · J.R. Martí-Vargas · P. Serna
ICITECH Institute for Concrete Science and Technology,
Universitat Politècnica de València, Camino de vera s/n Building 4N,
46022 Valencia, Spain
e-mail: aillator@upv.es

S.E. Arango
Cementos Argos S.A., Medellín, Colombia

E. García-Taengua
School of Civil Engineering, University of Leeds, Leeds LS2 9JT, UK

P. Serna et al. (eds.), *Creep Behaviour in Cracked Sections of Fibre Reinforced Concrete*, RILEM Bookseries 14,
DOI 10.1007/978-94-024-1001-3_16

1 Introduction

The knowledge of the concrete materials and properties is essential to better assess their structural applications, not only in terms of instantaneous responses but also in the long-term along the service life.

Creep is a term used to define the tendency of materials to develop increasing strains through time when under a sustained load. Commonly, the increasing strains are accounted for by means of creep coefficients that relate long-term strains with respect to short-term strains. The long-term strains of concrete have been reported to become multiple times larger than the initial instantaneous strain [1]; therefore, creep becomes an important factor to be considered.

Long-term strains can be beneficial to some types of structures because they can lead to stress redistributions that can limit the extent of cracking. At the same time, in the case of decreasing residual strength or significant strains the influence of creep may be unfavourable.

In many applications, fibres have been included in concrete to improve structural serviceability based on benefits in crack control [2]. The mechanical behaviour of fibre reinforced concrete (FRC) has been widely studied over the past decades [3], whereas the analysis of flexural creep behaviour of cracked FRC elements is a relatively new topic which has not been entirely researched yet. Important advances have been achieved regarding the residual strength characterization and applications of FRC [2]. However, there is no standardized method to assess such behaviour at this time, and some researchers are working on proposals [1, 4].

When considering combinations of reinforcing bars and fibres, studies have shown that the addition of steel fibres considerably reduces the time-dependent deflection and crack widening of reinforced concrete [5, 6]. However, in terms of structural design, recommendations and codes do not take into account the long-term behaviour under cracked conditions.

As the contribution of fibres to structural load-bearing capacity is based on the flexural response of FRC, and mainly in the cracked state, the capacity of the material to keep the crack opening values low enough is a topic of interest to be assessed [4]. Although residual strength parameters are applied in structural design, the knowledge of the long-term behaviour of FRC in cracked conditions is limited to a few reports on the subject [1–3, 5, 7–9].

The study of creep behaviour of cracked FRC constitutes a key point of interest mainly for cases in which fibres are the only reinforcement. In these cases, serviceability of the material will depend on its capacity to transfer the sustained stresses through the fibres and the stability of the cracks [5].

A previous study on cracked SFRC [10] has found low coefficients of creep at different long-term load levels for a 90-day period, but the type of fibre, the load level and the concrete strength significantly affected the creep behaviour. Besides, it is not possible to uncouple the behaviour of concrete from that of the fibres because of the post-cracking creep phenomena in SFRC is not caused by deformation of fibres, but rather by pull-out of the fibres from the matrix [5].

This paper studies the influence of fibre reinforcement on long-term behaviour of cracked SFRC by means of a creep test analysing not only creep coefficients but also Crack Opening Rates.

2 Experimental Program

2.1 Materials and Mix Design

Nine concretes were prepared based on two concrete base mix designs as presented in Table 1. Seven steel fibre reinforced concretes (SFRC) and two conventional reinforced concretes (RC) were included.

Concretes type I were conceived to simulate a concrete for structural pre-casting purposes with a compressive strength higher than 40 MPa, and concretes type II were designed to reproduce general purpose concretes with a 10 or 20 mm maximum aggregate size and a compressive strength ranging from 25 to 35 MPa.

For type I concretes a cement CEM I 52.5R type and a polycarboxylate-based high-range water reducing admixture were used, whereas for type II concretes the cement was CEM I 42.5R SR or CEM II/BM 42.5R. In those cases the water reducer was a poly-functional plasticiser. The dosage of plasticiser was adjusted for each concrete to obtain a 100 ± 20 mm slump. For all concretes, sand and coarse aggregates were crushed limestone.

2.2 Reinforcements

Variables related to fibre dosages and conventional reinforcements were selected to cover a wide range of post-peak flexural response of the SFRC elements with

Table 1 Base mixture proportions and concrete properties

Parameter	Concrete type	
	I	II
Compressive strength (f_c)	≥40 MPa	25-35 MPa
Cement type	CEM I 52.5R	CEM I 42.5R SR
		CEM II/B-M 42.5R
Cement amount	375 kg/m^3	325 kg/m^3
W/C	0.5	0.6
Fibre application	Structural-precast	Pavements
Fibre volume (V_f)	40 and 70 kg/m^3	40 kg/m^3
Maximum aggregate size (MAS)	10 mm	10 and 20 mm
Admixture	High-range water reducer	Poly-functional plasticizer

Table 2 Characteristics of fibres

Designation	Slenderness (λ)	Length (mm)	End type	Type
Dramix RC 80/35 BN	80	35	Hooked	Cold drawn
Dramix RC 80/50 BN	80	50	Hooked	Cold drawn
Dramix RC 65/40 BN	65	40	Hooked	Cold drawn
Dramix RL 45/50 BN	45	50	Hooked	Cold drawn
Fibrocev F-due 50/30	50	30	Straight	Cut sheet

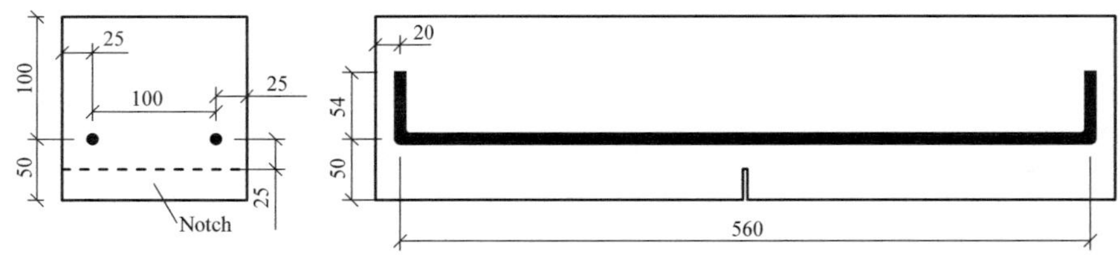

Fig. 1 Rebar reinforced specimens

concretes showing flexural strain hardening behaviour, concretes that keep the load bearing capacity practically constant (flat post-peak region) and concretes with steep losses of residual strength after first crack (softening behaviour). In all cases of fibre reinforcement steel fibres were used. Some characteristics of the steel fibres used are given in Table 2.

Concrete specimens without fibres as reinforcement, both for characterisation and creep tests, were reinforced simulating a conventional RC by means of two steel rebars positioned in a longitudinal way as shown in Fig. 1.

Table 3 summarises for all batches designations the reinforcement characteristics like fibre dosages, fibre brand and dimensions of conventional reinforcement.

Table 3 Reinforcement characteristics for all series

Concrete type	Designation	Reinforcement	Dimensions	Dosage (kg/m^3)	Maximum aggregate size
Type I	I-80/35-40-10	Fibres	RC 80/35 BN	40	10
	I-80/35-70-10	Fibres	RC 80/35 BN	70	10
	I-80/50-40-10	Fibres	RC 80/50 BN	40	10
	I-2Ø8	Steel rebars	8 mm bars (×2)		10
Type II	II-80/50-40-20	Fibres	RC 80/50 BN	40	20
	II-65/40-40-20	Fibres	RC 65/40 BN	40	20
	II-45/50-40-20	Fibres	RL 45/50 BN	40	20
	II-50/30-40-10	Fibres	F-DUE 50/30	40	10
	II-2Ø6	Steel rebars	6 mm bars (×2)		20

2.3 Test Specimens

Both FRC and RC specimens were cast in 150 × 150 × 600 mm size moulds, cured and prepared by following the recommendations of the EN 14651:2007 standard [11] for flexural tests.

Compaction was carried out by external vibration in order to avoid any differential effect on the fibre orientation along the control section among the tested concretes.

Prior to the test, specimens were rotated over 90° around their longitudinal axis and then notched by sawing through the width of the test specimen at mid span. The notch depth was 25 mm.

All concretes were characterised in terms of compression and their flexural residual strength.

2.4 Creep Test Variables

The experimental campaign included a total of 37 SFRC and 6 RC specimens which were successfully tested according the flexural creep test methodology detailed in the next section. Table 4 shows the tested combinations.

In addition to the variables of mix design of concrete, there are two significant variables related to the creep test: the nominal pre-crack opening level ($CMOD_{pn}$) and the stress level (I_c) sustained during creep test. In this work, the $CMOD_{pn}$ takes values of 0.5 and 1.5 mm which represent the first and second level where the corresponding flexural residual strength (f_{R1} and f_{R2}) are calculated following the EN 14651:2007 standard [11]. Stress levels or I_c values were defined aiming at testing the concretes from a service load level (60 % of $f_{R,p}$) to a relatively high load level (80 % of $f_{R,p}$) and at an exceptional load level (95 % of $f_{R,p}$).

Table 4 Creep test program

Designation	Nominal stress level (I_n, %)	Pre-crack ($CMOD_{pn}$, mm)	Tested specimens
I 80/35-40-10	60, 80	0.5	6
	80	1.5	3
I 80/35-70-10	60, 80, 95	0.5	9
I 80/50-40-10	80	0.5	2
I 2Ø8-10	80	0.5	3
II 80/50-40-20	80	0.5	3
II 65/40-40-20	60	0.5	3
II 45/50-40-20	80	0.5	6
II 50/30-40-10	60, 80	0.5	5
II 2Ø6-20	80	0.5	3

3 Flexural Creep Test Methodology

Creep tests were performed by applying a constant bending load on 150 × 150 × 600 mm notched pre-cracked specimens. The applied load and the crack opening evolution during the test were monitored. Figure 2 shows a general view of the creep test specimens. Full explanation of the creep test methodology, details about frame and components and dimensions specifications can be found in Arango's PhD. Thesis [10] (in Spanish) and further publications [4].

A creep test consists basically on three main stages: *pre-cracking stage*: specimen pre-cracking up to a nominal crack opening value "$CMOD_{pn}$"; *creep test stage*: test on creep of pre-cracked specimens subjected to a creep stress level "I_c"; and *final bending test stage*: final test in flexure to failure on specimen after creep process.

As a multiple specimens setup in column was adopted, some aspects of the loading configuration required to be modified to ensure the column stability. In particular, the three point bending test proposed in [11] was adapted to a four point bending test as shown in Fig. 3. This change make it possible also to measure

Fig. 2 General view of creep test specimens (overview *-left-* and column detail *-right-*)

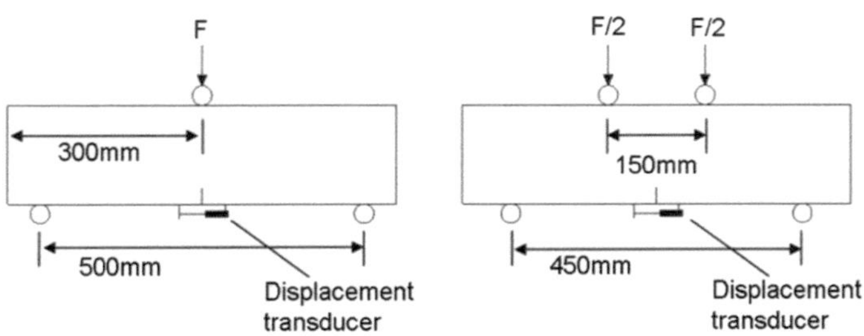

Fig. 3 Flexural test diagrams: EN 14651:2007 (*left*) and creep test proposal (*right*)

compressive strains by means of strain gauge located centred at the top face of the specimens. Accordingly, some changes in stress formulation were also realised, since the load configuration was changed.

3.1 Test Specimen Pre-cracking

Test specimens pre-cracking is done in a 4 PBT set up at the loading rate proposed in EN 14651:2007 [11], but when the nominal crack opening value $CMOD_{pn}$ is reached the test is stopped and the specimen unloaded. The load–crack opening evolution during the loading, unloading and recovery process is registered. Figure 4 presents an idealization of the pre-cracking process which include the following main parameters: first crack stress (f_L), maximum actual crack opening value on pre-cracking process ($CMOD_p$), stress at $CMOD_p$ ($f_{R,p}$) and residual crack opening value after pre-cracking process ($CMOD_{pr}$).

3.2 Creep Test

After the pre-cracking stage, the test specimens are placed inside a chamber with controlled environmental conditions where the creep frames are located. The temperature of the chamber is 20 °C and the relative humidity is 50 %.

Each frame is able to test simultaneously a column of three test specimens (see Fig. 2). Once the specimens were ready, the creep test stage starts when the load is applied. The sustained nominal stress level or nominal index of stress in creep (I_n) is

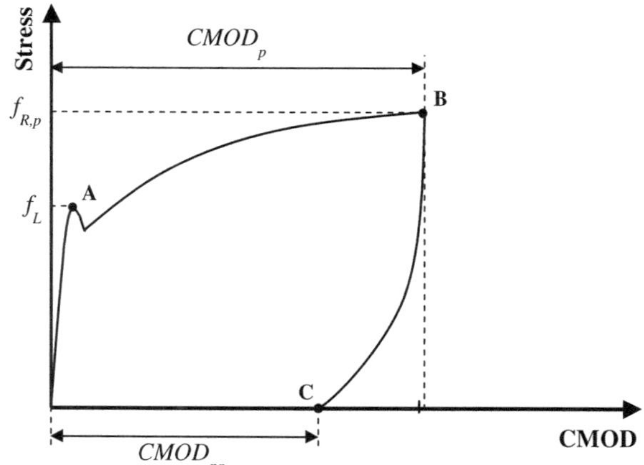

Fig. 4 Pre-cracking stage parameters definition

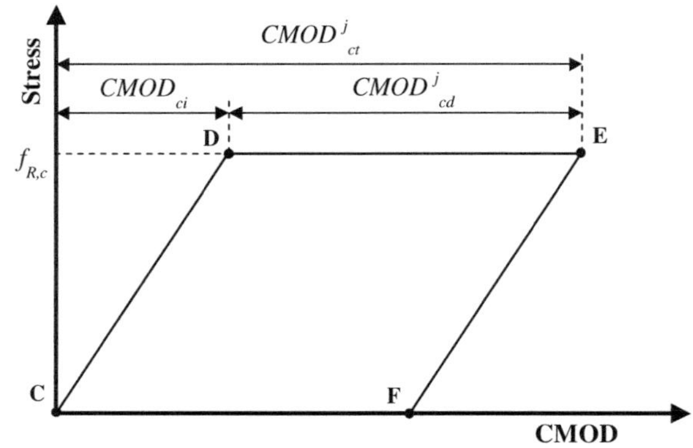

Fig. 5 Creep tests phase parameters definition

defined as a percentage of the residual strength at $CMOD_{pn}$ ($f_{R,p}$). Due to the multiple specimen set-up, the load applied is not the same for all the specimens. Therefore, as their residual strength is also variable, the real stress level or real index of stress in creep (I_c) for each specimen is calculated and defined as the ratio between the sustained stress during creep phase ($f_{R,c}$) and the residual strength at $CMOD_{pn}$ ($f_{R,p}$) by the Eq. (1).

$$I_c = f_{R,c}/f_{R,p}\ (\%) \tag{1}$$

The test continues without interruption until it is decided to unload the specimens. Once specimens are unloaded they remain in the creep frame and the recovery of deformations is recorded for two weeks. Figure 5 represents an idealisation of the creep stage, where it can be identified some parameters: the sustained stress during creep phase ($f_{R,c}$), the instantaneous crack opening deformation ($CMOD_{ci}$), the deferred crack opening deformations at different ages ($CMOD^j_{cd}$) and the total deformations during creep test ($CMOD^j_{ct}$) as the addition of instantaneous deformations and deferred deformations.

3.3 Post-creep Flexural Bending Test to Failure

When the creep test is over, the test specimens are subjected to post-creep flexural tests to failure, following the same methodology exposed in 3.1 reaching crack opening values higher than 4 mm.

By assembling the experimental results of the previously mentioned stages and plotting Stress-CMOD, the evolution of the measured test parameters is represented considering the complete test to establish a common origin for crack opening values for all test stages.

4 Test Results and Analysis

4.1 Characterization Tests

Flexural characterisation tests were performed to all series of FRC following the EN 14651:2007 standard [11]. Figure 6 shows the Load–CMOD curves obtained in these tests. As it can be observed, a wide spectrum of behaviours comprising hardening, flat and softening response was covered.

Table 5 resumes both the compressive strength results for each series and the residual strengths at Limit of Proportionality (LOP) and for CMOD values of 0.5 and 2.5 mm. The coefficient of variation is given as informative of the scatter of results.

4.2 Creep Tests

CMOD deformations were registered during flexural creep tests. As reference for both types of concretes, Fig. 7 shows the deferred deformation evolution with time of one specimen of each series. Notice that, for concretes type II, instantaneous deformations are smaller than for concretes type I and ranges from 0.2 to 0.5 mm.

In addition, a number of specimens (40 % of them) developed a sudden increase of crack opening deformations during creep stage. A visual example of these sudden increases can be observed in Fig. 7.

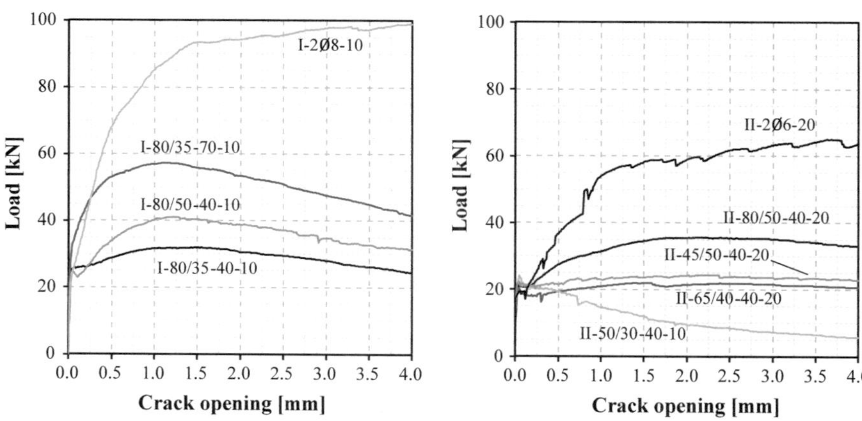

Fig. 6 Flexural behaviour of concretes type I (*left*) and II (*right*)

Table 5 Characterisation tests results (compressive strength and residual strengths)

Concrete	Comp. strength		Residual flexural strength (MPa) & CoV (%)					
	f_c		f_L		f_{R1}		f_{R3}	
80/35-40-10	52.7	(10.0)	4.94	(6.5)	5.53	(6.0)	5.66	(6.1)
80/35-70-10	56.1	(4.9)	6.38	(4.9)	10.12	(10.6)	9.78	(9.8)
80/50-40-10	57.1	(4.3)	4.53	(4.9)	6.14	(11.7)	6.79	(5.4)
2Ø8-10	58.3	(5.3)	4.80		13.10		18.36	
80/50-40-20	39.0	(1.3)	3.93	(2.7)	5.01	(8.6)	6.48	(9.6)
65/40-40-20	29.4	(6.6)	4.03	(2.8)	3.72	(14.0)	3.79	(17.0)
45/50-40-20	38.8	(6.6)	4.27	(4.2)	4.09	(22.9)	4.58	(10.4)
50/30-40-10	41.3	(6.0)	4.68	(3.3)	3.69	(11.6)	1.63	(20.3)
2Ø6-20	36.4	(11.2)	3.68		7.03		11.85	

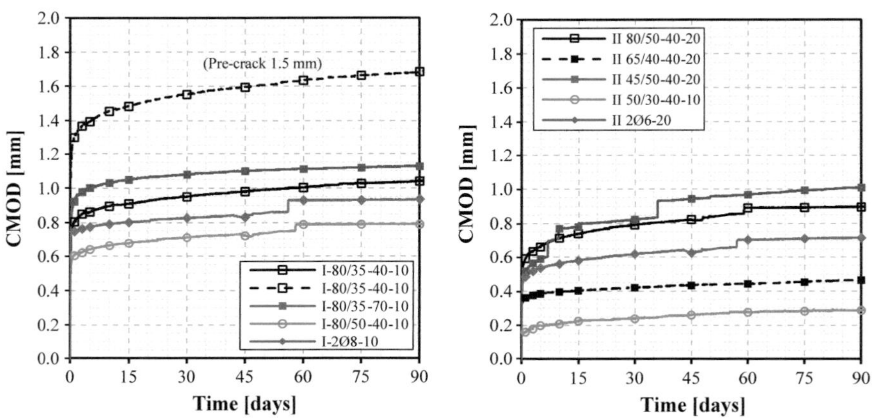

Fig. 7 CMOD–time curves for concretes type I (*left*) and II (*right*)

In case of concretes type I, these sudden increases were produced in some specimens of series reinforced with fibres with slenderness (λ) of 80 and stress level $I_c \geq 90$ %. For concretes type II, SFRC with fibres with slenderness (λ) of 45 and 50 developed sudden CMOD deformations when $I_c \geq 70$ %. The magnitude of these sudden increases of crack opening during creep tests varied from 0.012 to 0.559 mm, what supposed a percentage of 2–42 % of deferred CMOD deformation at 90 days of the corresponding specimen.

These sudden increases were considered as deferred deformations for the time lapses when they occurred.

4.3 Results Analysis

The deferred Crack Mouth Opening Displacements (CMOD) are needed to calculate some creep parameters to analyse the long-term behaviour. As time reference, three specific lapses of time were selected: 14, 30 and 90 days. The parameters worked out to evaluate creep behaviour were the creep coefficients and the Crack Opening Rates (COR).

4.3.1 Creep Coefficients

The creep coefficient (φ) is defined as the ratio between the deferred deformation and instantaneous deformation. Deferred deformations were obtained for the selected time lapses to check the evolution of creep coefficients along time. Referring the creep coefficient to the creep stage, the instantaneous deformations occurred during the loading phase of creep tests. In this way, the instantaneous deformation was recorded just when the creep stress level I_c was reached (point D in Fig. 5). Therefore, the creep coefficient referred to creep stage at time j (φ_c^j) can be calculated by means of Eq. (2):

$$\varphi_c^j = \mathrm{CMOD}_{\mathrm{cd}}^{\mathrm{j}} / \mathrm{CMOD}_{\mathrm{ci}} \tag{2}$$

where $\mathrm{CMOD}_{\mathrm{cd}}^{\mathrm{j}}$ is the deferred crack opening at time j and $\mathrm{CMOD}_{\mathrm{ci}}$ is the instantaneous crack opening occurred during the loading phase of creep tests (see Fig. 5).

Since the specimens were previously pre-cracked, there was a previous crack opening deformation before the creep test. Therefore, a new creep coefficient referred to the origin of deformations at time j (φ_o^j) can be calculated by means of Eq. (3):

$$\varphi_o^j = \mathrm{CMOD}_{\mathrm{cd}}^{\mathrm{j}} / \mathrm{CMOD}_{\mathrm{ci}}^{\mathrm{o}} \tag{3}$$

where $\mathrm{CMOD}_{\mathrm{ci}}^{\mathrm{o}}$ is the total deformation referred to the origin obtained by means of Eq. (4):

$$\mathrm{CMOD}_{\mathrm{ci}}^{\mathrm{o}} = \mathrm{CMOD}_{\mathrm{pr}} + \mathrm{CMOD}_{\mathrm{ci}} \tag{4}$$

where $\mathrm{CMOD}_{\mathrm{pr}}$ is the residual crack opening after the pre-cracking test (see Fig. 4) and $\mathrm{CMOD}_{\mathrm{ci}}$ is the instantaneous deformations occurred during the loading phase of creep tests (see Fig. 5).

Both creep coefficients, referred to creep stage and referred to origin, where obtained for all specimens and compared. Figure 8 shows the creep coefficients results at 90 days for all creep stress levels I_c, referred to creep stage and to origin of deformations. Notice that some specimens with $I_c \geq 80$ % develop high creep

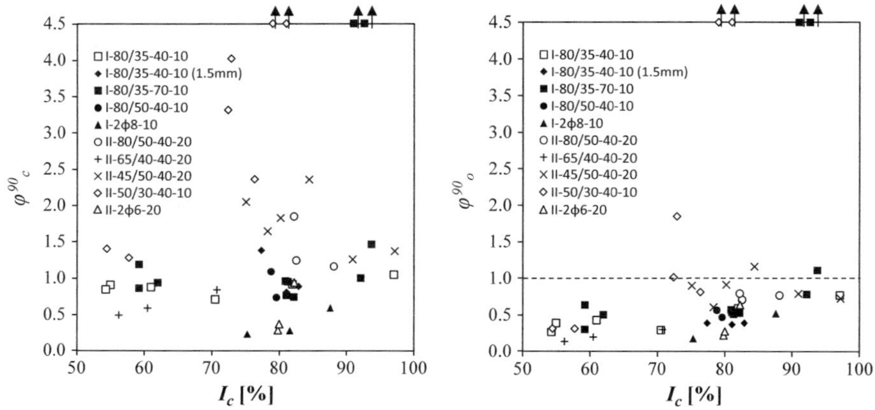

Fig. 8 Creep coefficients at 90 days referred to creep stage (φ_c^{90}) *-left-* and origin of deformations (φ_o^{90}) -right-

coefficients values. This fact was given due to sudden increases of CMOD deformations during creep test.

In case of SFRCs, the highest values of creep coefficients were given for those concretes reinforced with fibres with less slenderness (λ = 45–50) and a high length (l = 50 mm). In those cases when low $I_c \leq 70$ % is given, creep coefficients referred to origin of deformations for all concretes remains always below $\varphi = 1$. The Model Code [12] stablish this limit ($\varphi = 1$) as the reference creep coefficient value for concretes. For higher stress levels, creep coefficients referred to origin of deformations are closer to 1 but most of them still remain below $\varphi = 1$. Those specimens that suffered sudden increases of deformations during creep stage are clearly over the limit of $\varphi = 1$.

On the other hand, for conventional RCs, both types of concretes (I and II) obtain low values for both creep coefficients even with a high I_c. In case of creep coefficients referred to origin of deformations, the values are below the MC recommendation. Due to the high residual strength of RC specimens, this behaviour may be influenced by creep deformations in compression zone of the specimens. It should be desirable to reach the way to distinguish between *tensile creep* in cracked zone and *compressive creep* in compression zone of the specimen in a flexural creep test.

4.3.2 Crack Opening Rate (COR)

The Crack Opening Rate or COR is a parameter that helps to evaluate the average velocity of deferred deformations between two time lapses. This COR parameter is obtained by means of Eq. (5):

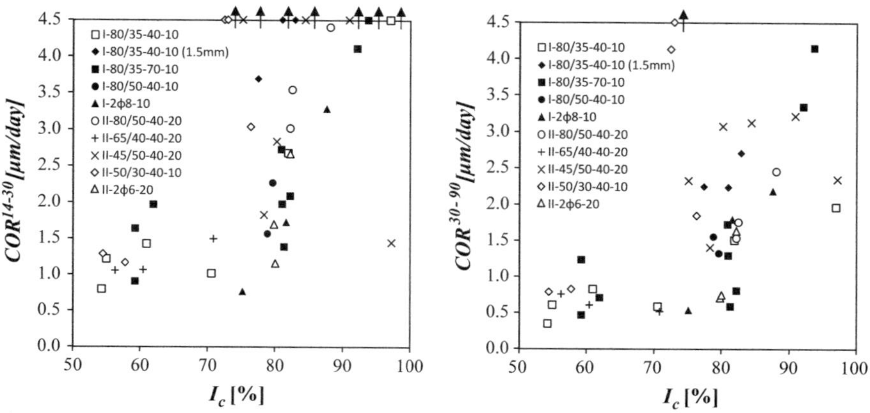

Fig. 9 COR^{14-30} (*left*) and COR^{30-90} (*right*) for all tested concretes

$$\mathrm{COR}^{j-k} = (\mathrm{CMOD}^{k}_{cd} - \mathrm{CMOD}^{j}_{cd})/(k - j) \tag{5}$$

where *j* and *k* are two different time lapses in days. This parameter is expressed in µm/day units. In this work, three lapses of time of creep test were analysed: from 0 to 14 days (2.74–59.72 µm/day), from 14 to 30 days (0.76–46.26 µm/day) and from 30 to 90 days (0.34–6.43 µm/day). Figure 9 shows the COR trends in two lapses of time: 14–30 and 30–90 days.

A global trend to reduce velocity with time can be observed. In case of FRCs type I, the minimum values were obtained for $I_c \leq 80$ % whereas for FRCs type II, minimum values were obtained for $I_c \leq 70$ %. In both concretes, lower values of COR were detected in case of conventional rebar reinforcement and SFRCs with low I_c.

Although there is a trend to reduce velocity along time, the COR values are quite high in general, especially is cases of high stress level I_c. These high COR values are not suitable since it may turn into a local or global failure during life time of the FRC element.

5 Conclusions

The influence of concrete matrix and fibre reinforcement on long-term behaviour of cracked concrete was analysed in this work. Concretes incorporating steel fibres and standard rebar reinforcement were studied. Creep coefficients referred to creep stage and referred to origin were obtained. Moreover, COR parameters were also calculated and analysed. The main conclusions obtained are:

- SFRC specimens with softening post-crack behaviour, develop more deferred deformations and higher risk of failure during creep test.

- A non negligible number of specimens (18 out of 45) developed sudden increases of crack opening deformations at high stress levels ($I_c > 75$ %).
- All rupture cases occurred during the loading phase in the creep test frames or a few minutes later once the stress level for creep tests was achieved and in cases of high stress level.
- Creep coefficients show significant dependence of analysed variables.
- For low stress levels ($I_c < 80$ %), creep coefficients remain at low values and regular.
- High values of creep coefficients correspond to high values of crack opening deformations.
- The highest values of creep coefficients were given for those concretes reinforced with fibres with less slenderness ($\lambda = 45$–50) and a high length of fibres (l = 50 mm).
- In those cases when low I_c is given and in most cases of high stress level, creep coefficients referred to origin of deformations for all concretes types remains always below the Model Code recommended value.
- The results of this experimental campaign show that creep in SFRC specimens may be similar than in a conventional RC. The use of fibres, even at low contents, is a good strategy to control flexural creep in the cracked state.
- Although a global trend to reduce velocity along time is observed, the COR values are quite high in general, especially in the cases of high stress level I_c. These high COR values are not suitable since it may turn into a local or global failure during life time of the FRC element.
- In order to reach the way to distinguish between tensile creep in cracked zone and compressive creep in compression zone in a flexural creep test, more studies are needed.

Acknowledgments The authors of this work wish to thank the Spanish Ministry of Economy and Competitiveness and the European Regional Development Fund, for the following funded projects "BIA2012-35776 (FISNE)".

References

1. Babafemi, A.J., Boshoff, W.P.: Testing and modelling the creep of cracked macro-synthetic fibre reinforced concrete (MSFRC) under flexural loading. Mater. Struct. doi:10.1617/s11527-016-0795-7, published online 14 January 2016
2. Zerbino, R., Monetti, D.H., Giaccio, G.: Creep behaviour of cracked steel and macro-synthetic fibre reinforced concrete. Mater. Struct. doi:10.1617/s11527-016-0795-7, published online 12 October 2015
3. Serna, P., Martí-Vargas, J.R., Bossio, M.E., Zerbino, R.: Creep and residual properties of cracked macro-synthetic fibre reinforced concretes. Mag. Concre. Res. **68**(4), 197–207 (2016)
4. Arango, S., Serna, P., Martí-Vargas, J.R., García-Taengua, E.: A test method to characterize flexural creep behaviour of pre-cracked FRC specimens. Exp. Mech. **52**(8), 1067–1078 (2012)

5. Zerbino, R.L., Barragan, B.E.: Long-term behavior of cracked steel fiber-reinforced concrete beams under sustained loading. ACI Mater. J. **109**(2), 215–224 (2012)
6. Vasanelli, E., Micelli, M., Aiello, M.A., Plizzari, G.: Long term behaviour of FRC flexural beams under sustained load. Eng. Struct. **56**, 1858–1867 (2013)
7. Bernard, E.S.: Influence of fiber type on creep deformation of cracked fiber-reinforced shotcrete panels. ACI Mater. J. **107**(5), 474–480 (2010)
8. Zhao, G., di Prisco, M., Vandewalle, L.: Experimental investigation on uniaxial tensile creep behaviour of cracked steel fiber reinforced concrete. Mater. Struct. **48**(10), 3173–3185 (2015)
9. García-Taengua, E., Arango, S., Martí-Vargas, J.R., Serna, P.: Flexural creep of steel fiber reinforced concrete in the cracked state. Constr. Build. Mater. **65**, 321–329 (2014)
10. Arango, S.E.: Fluencia a Flexión del Hormigón Reforzado con Fibras de Acero (SFRC) en Estado Fisurado. Doctoral Thesis, Universitat Politècnica de València, Spain (2010)
11. European Committee for Standardization: EN14651:2007 Test Method for Metallic Fibered Concrete—Measuring the Flexural Tensile Strength (Limit of Proportionality (LOP), Residual). Brussels, Belgium (2007)
12. Fédération Internationale du Béton. fib Model Code for Concrete Structures 2010. Berlin, Germany (2013)

Creeping Effect of SFRC Elements Under Specific Type of Long Term Loading

Darko Nakov, Goran Markovski, Toni Arangjelovski and Peter Mark

Abstract Time-dependent properties of any type of concrete is of crucial importance to use it as a structural material. This type of research needs serious financial resources, long-term occupation of a laboratory and a lot of time for testing. Several studies have been conducted on long-term behaviour of SFRC beams under sustained loads, while studies which include the effect of repeated variable load are uncommon. In order to study the influence of different fibre dosages and the influence of continuously repeated variable load on the creeping effects of SFRC, an experiment was carried out at the FCE–Skopje. The experiment consisted of 24 RC beams (C30/37) with cross section 15/28 cm and a total length of l = 300 cm. According to the type of material they were divided in three series: without fibres; with 30 kg/m^3 and with 60 kg/m^3 hooked-end steel fibres. Regarding the loading history, the beams were divided into four groups, by which the 3rd and 4th group were dealing with long term load. The obtained experimental results are presented and discussed in terms of time-dependent: deflections and cracks width. All results indicate the positive influence of additional reinforcing of RC beams with steel fibres.

Keywords Steel fiber reinforced concrete · Experiment · Creep · Deflections · Crack widths

1 Introduction

Having in mind the positive influence of steel fibres on the general behaviour of structures, they are very often added to the standard or special concrete mixtures in some most recent applications, like the CCTV (China Central Television) tower in

D. Nakov (✉) · G. Markovski · T. Arangjelovski
University "Ss. Cyril and Methodius", FCE-Skopje, Skopje, Republic of Macedonia
e-mail: nakov@gf.ukim.edu.mk

P. Mark
Ruhr-University Bochum, Bochum, Germany

P. Serna et al. (eds.), *Creep Behaviour in Cracked Sections of Fibre Reinforced Concrete*, RILEM Bookseries 14,
DOI 10.1007/978-94-024-1001-3_17

Beijing, open roof building "Maison de l'ecriture" in Switzerland or the Oceanographic park in Valencia. But, all these special structures needs special attention regarding their time-dependent behaviour.

Deflections and crack widths predicted on the basis of short-term tests do not provide satisfactory results for verification in the serviceability limit states. That is why long term experiments of steel fibre reinforced concrete elements under sustained loads are very important. Long term sustained load causes a significant increase of deflections and crack widths. In addition, long term repeated variable load causes additional increase. This was proven by the experimental and theoretical research done in the last 12 years at the Faculty of Civil Engineering—Skopje. Following the original idea of Atanasovski S. [1], the effect of realistic load histories on the creep and structural behaviour was studied in several research projects. Markovski G. studied the influence of variable loads to the time-dependent behaviour of prestressed concrete elements [2], while Arangjelovski T. studied the time-dependent behaviour of reinforced high-strength concrete elements under the action of variable loads [3].

Up to now, there are only few published research studies dealing with long term behaviour of SFRC beams. Tan et al. in 1994 studied the behaviour of SFRC beams under sustained load as a one year study [4] and later the same research was continued as a ten year study of the deflections and crack widths under sustained load by Tan and Saha in 2005 [5]. Another research by Vasanelli et al. in 2012 [6] is dealing with the influence of long term sustained loading on the cracking behaviour and consequently on the structural durability.

This paper presents experimental results from the research project [7] of the first author related to the influence of different fibre dosages and the influence of continuously repeated variable load on the creeping effects of steel fibre reinforced concrete.

2 Experimental Program

The experiment was carried out at the University "Ss. Cyril and Methodius", Faculty of Civil Engineering-Skopje, Republic of Macedonia, in the period October 2011 to February 2013. It involved testing of 24 full scale beams constructed from reinforced concrete and steel fibre reinforced concrete with additional reinforcement. The beams had a cross section proportioned 15/28 cm and a total length of l = 300 cm, Fig. 1. Together with each series of beams, control specimens were cast in order to test the compressive strength, flexural tensile strength, splitting tensile strength, elastic modulus and deformations due to creep and shrinkage. In addition to the tests on mechanical and time-dependent properties of concrete, the used reinforcement was also tested.

All 24 beams were manufactured with concrete class C30/37. According to the used type of material, they were divided into three series:

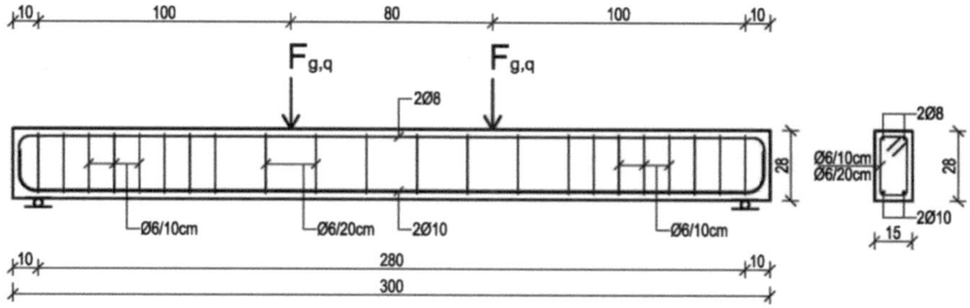

Fig. 1 Geometry, reinforcement and loading scheme of full scale beams

- Series A, reinforced concrete (C30/37);
- Series B, SFRC with 30 kg/m^3 steel fibres and additional reinforcement (C30/37 FL1.5/1.5);
- Series C, SFRC with 60 kg/m^3 steel fibres and additional reinforcement (C30/37 FL2.5/2.0).

The beams constructed of reinforced concrete were used for comparison with the beams constructed of steel fibre reinforced concrete. In each series, the plain reinforcement was kept the same. The longitudinal reinforcement was ribbed and of RA 400/500-2 quality, while the shear reinforcement was smooth, with GA 240/360 quality. Reinforcement 2Ø10, 2Ø8 and Ø6/10/20 cm was used as tension, compression and shear reinforcement, respectively.

The used steel fibres were hooked-end HE1/50, produced of cold-drawn wire, manufactured by Arcelor Mittal, with a diameter of 1 mm, length of 50 mm and tensile strength of 1100 N/mm^2.

The experimental program is presented in detail in Table 1.

The mixture proportioning (Table 2) was done so that it was the same for the three types of concrete. According to many recommendations in the up to date literature, the slump of the concrete without fibres was 120 mm. Since fibres decrease workability, the slump was decreased to 75 and 50 mm with addition of 30 and 60 kg/m^3.

Regarding the loading history, the beams were divided into four groups (Fig. 2):

1. The beams from all three series from group “1” (A_1, B_1, C_1) were tested under short term ultimate load at the age of concrete of 40 days. With these testing, relevant dependences had to be found for this age of concrete and the behaviour of the reinforced concrete and two types of steel fiber reinforced concretes had to be compared.
2. The beams from all three series from group “2” (A_2, B_2, C_2) were tested also under short term ultimate load, but at the age of concrete of 400 days. This testing was performed in order to find out the influence of the age of concrete on the behaviour of the beams.
3. The beams from group “3” (A_3, B_3, C_3) were pre-cracked with permanent and variable load “g + q”, and afterwards, a long term permanent load with intensity

Table 1 Experimental program

Series	Group	Number of elements	Type of concrete	Steel fibres (kg/m^3)	Tensile reinforcement μ (%)	Type of long term load	Time of ultimate load testing (days)
A 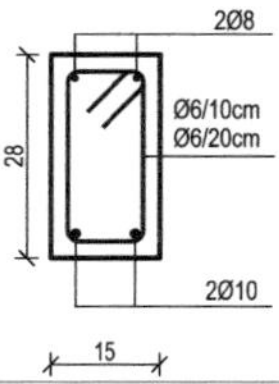	1	2	C30/37	0	0.37	/	t = 40
	2	2	C30/37	0	0.37	/	t = 400
	3	2	C30/37	0	0.37	"g" *	t = 400
	4	2	C30/37	0	0.37	"g ± q" (Δt_{g+q}=8 h)	t = 400
B 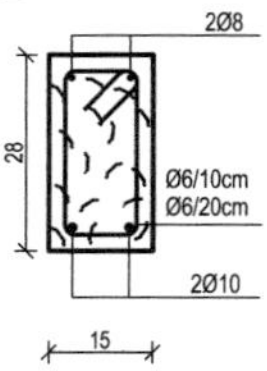	1	2	C30/37 FL 1.5/1.5	30	0.37	/	t = 40
	2	2	C30/37 FL 1.5/1.5	30	0.37	/	t = 400
	3	2	C30/37 FL 1.5/1.5	30	0.37	"g" *	t = 400
	4	2	C30/37 FL 1.5/1.5	30	0.37	"g ± q" (Δt_{g+q}=8 h)	t = 400
C 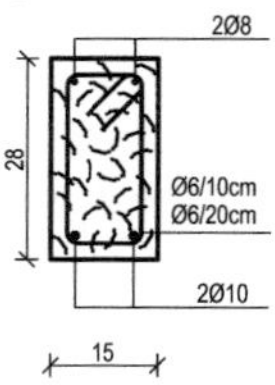	1	2	C30/37 FL 2.5/2.0	60	0.37	/	t = 40
	2	2	C30/37 FL 2.5/2.0	60	0.37	/	t = 400
	3	2	C30/37 FL 2.5/2.0	60	0.37	"g" *	t = 400
	4	2	C30/37 FL 2.5/2.0	60	0.37	"g ± q" (Δt_{g+q}=8 h)	t = 400

Table 2 Mixture proportions for the three concrete types

Mixture proportions	(kg/m^3)
Cement CEM II/A-M 42.5 N	410
Water	215
Water/Cement ratio, w/c	0.524
Aggregate:	
0–4 mm (river sand), 50 %	875
4–8 mm (limestone), 20 %	350
8–16 mm (limestone), 30 %	525
Fibres:	
C30/37	0
C30/37 FL 1.5/1.5	30
C30/37 FL 2.5/2.0	60

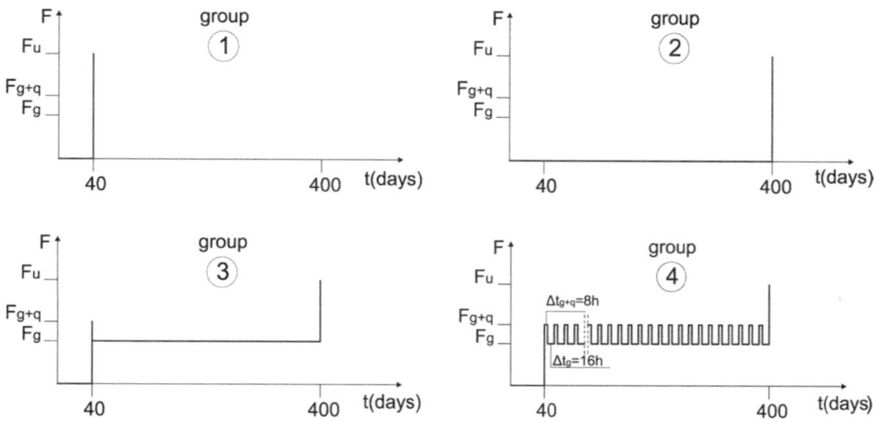

Fig. 2 Loading history for the beams of the four groups

"g" was applied at the age of concrete of 40 days and was held up to 400 days, when a short term ultimate load testing was performed. In the meantime, the strains, deformations and crack widths were measured.

4. On the beams from group "4" (A_4, B_4, C_4), a long term permanent load with intensity "g" was applied at the age of concrete of 40 days and was held for a year as a long term load. On the fortieth day, repeated variable load "±q" was also applied in an interval of 8 h +q and 16 h −q, for a year. This means that the beams were loaded additionally with load "q" for 8 h every day, whereas the strains, deformations and crack widths were measured. After 8 h, the beams were unloaded from load "q" and all measurements were performed again.

The beams and control specimens were cured for 8 days and then they were transported to the Laboratory at the Faculty of Civil Engineering—Skopje, where they were kept under almost constant temperature ranging from 15 to 24 °C with an average of 19.5 °C and constant relative ambient humidity ranging from 58 to 62 % with an average of 60.2 %, which was regulated with special humidifiers and dehumidifiers, presented in Fig. 3.

On the 12 full scale beams within all three series from group 1 and 2, short term ultimate load test was performed in 29 load steps. A data acquisition system from

Fig. 3 Ambient conditions in the laboratory of the faculty of civil engineering—Skopje

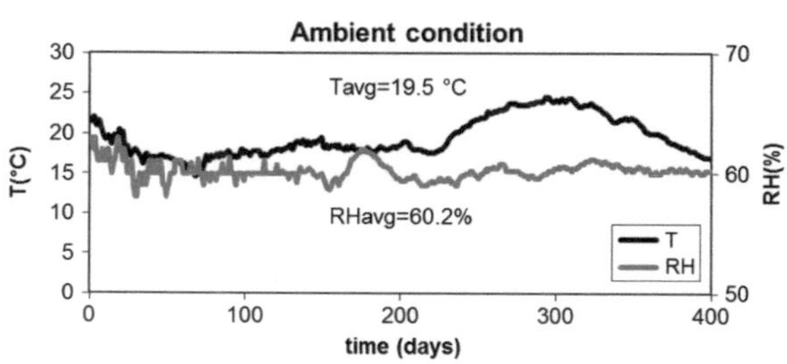

Hottinger Baldvin-HBM, Germany was used for recording the force, the middle deflection U3, the strains in the compression and tensile reinforcement (R1–R4) and the bottom and top of the concrete (C1–C4), with f = 1 Hz. The load cell and the LVDT as well as the strain gages were a product of Kyowa, Japan. In each step, the strains in the concrete (D1–D15), in the middle section of the beam through the thickness as well as on the top of the beam, were measured by a mechanical deflection meter, type Hugenberger, Switzerland, with a base of 250 mm. The mechanical measurement of the deflections was done at 5 points through the length of the beam and 2 points over the supports by using deflection meters produced by Stopani, Italy. The crack widths were also measured in each load step, in the region with constant moment, by use of a crack microscope—product of Controls, Italy. After 360 days of observing of their behaviour under the effect of long term loads, an ultimate load test was performed in the same manner as described above. The positions of the measurement points are presented in Fig. 4.

The long term load, which consists of permanent sustained load "g" and repeated variable load "q", was applied by 12 gravitation levers, which enabled an increase of the load for 13 times. The permanent load acts all the time, while the variable load was applied and removed each day by secondary hand gravitation levers.

The bending moments are as follows: from self-weight of the beam, M_{sw} = 1 kN m, from permanent load "g", M_g = 5.0 kN m, from variable load "q", M_q = 3.1 kN m, from self-weight, permanent and variable load (service) M_{sw+g+q} = 9.1 kN m. The bending crack moment was M_{cr} = 6.1 kN m, while the ultimate bending moment M_d = 15.6 kN m. The intensity of the load was chosen so that the M_{cr} is bigger than M_{sw+g} and smaller than M_{sw+g+q}. The permanent load is 0.39 times the flexural strength, while the service load is 0.58 times the flexural strength of the beam without fibres.

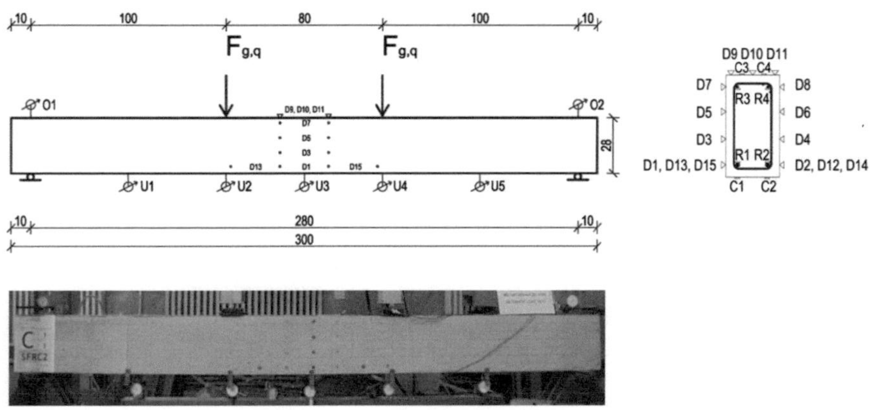

Fig. 4 Positions of measurement points of full scale beams

3 Selected Experimental Results

The results from the long term loading of group 3 and 4 are presented in terms of time-dependent: deflections and cracks width.

3.1 Time-Dependent Deflections

The beams from group 3 were first loaded with permanent load "g". In order to induce cracks and activate the fibres, the beams were precracked with load "g + q" and than the load was returned to the level of permanent load which was acting in the period of 360 days. In Figs. 5 and 6, total and long term deflections are presented as an average value of two beams for each concrete type.

From the figures can be noticed that the instantaneous, total and long term deflection decreased with the addition of fibres.

Instantaneous deflection after precracking with the load "g + q", at the level of permanent load "g", for the concrete residual strength classes FL 1.5/1.5 and FL 2.5/2.0, decreased for 19.6 % and 32.5 %, respectively.

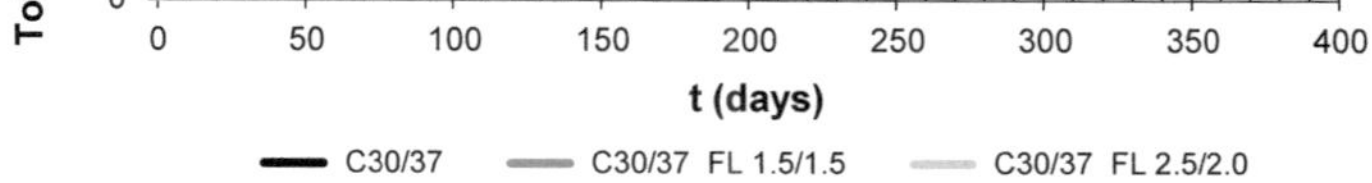

Fig. 5 Total time-dependent deflection for the beams from group 3

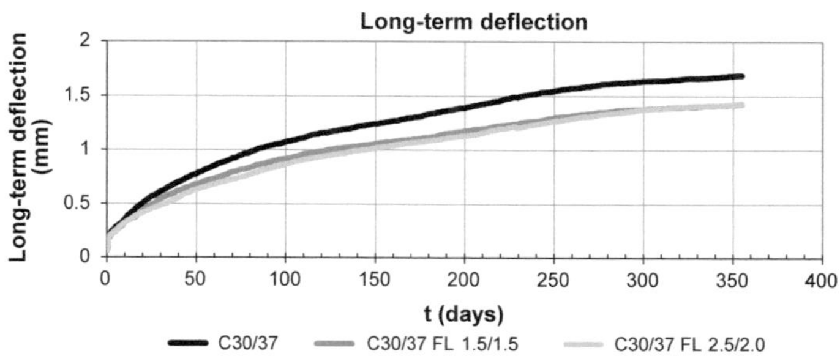

Fig. 6 Long term deflection for the beams from group 3

Total deflection after being subjected to permanent sustained load "g" for 360 days, for the concrete residual strength classes FL 1.5/1.5 and FL 2.5/2.0 decreased for 17.8 % and 24.5 %, respectively.

Long term deflection, which is the total minus instantaneous deflection after precracking, at the level of permanent load "g", for the concrete residual strength classes FL 1.5/1.5 and FL 2.5/2.0 decreased for 15.6 and 15.3 %, respectively. It can be noticed that both concretes with the addition of fibres decreased the deflections for about the same percentage.

The beams from group 4 were first loaded with permanent load "g" and after the measurements were performed they were loaded additionally with variable load "q". The variable load "q" was kept for 8 h and removed in the next 16 h. This was repeated every day in the period of 360 days. Before and after each application or removal of the variable load, which means four times each day, all the necessary measurements were performed.

In Figs. 7 and 8, total and long term deflections are presented as an average value of two beams for each concrete type.

From the previous figures, it can be noticed that the instantaneous, total and long term deflection decreased with the addition of fibres.

Instantaneous deflection at the level of service load "g + q", for the concrete residual strength classes FL 1.5/1.5 and FL 2.5/2.0 decreased for 7.6 and 22.8 %, respectively.

Total deflection after being subjected to permanent sustained and repeated variable load "g ± q" for 360 days, decreased for 14.4 and 22.5 % at the level of permanent load, respectively for residual class FL 1.5/1.5 and FL 2.5/2.0.

Long term deflection after 360 days under load, at the level of permanent load "g", which is the total deflection minus instantaneous deflection at the level of service load "g + q", also decreased for 19.2 and 22.2 %, respectively. It can be noticed that both concretes decreased the deflections for about the same percentage with the addition of fibres and increasing the residual strength.

The comparison between the deflections of the beams from group 3 and 4 (Table 3) with a resulting ratio, which were subjected to different load histories, is

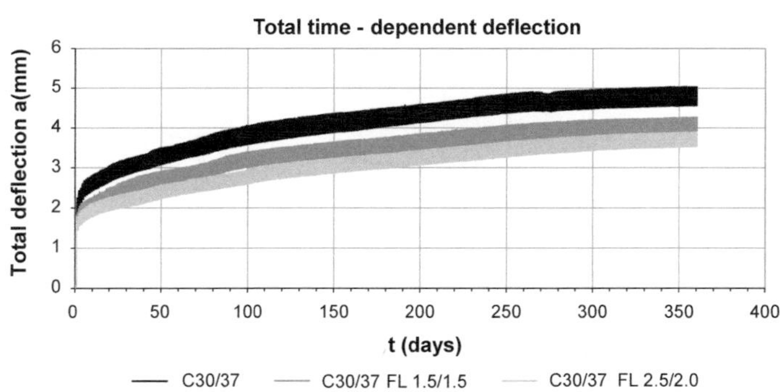

Fig. 7 Total time-dependent deflection for the beams from group 4

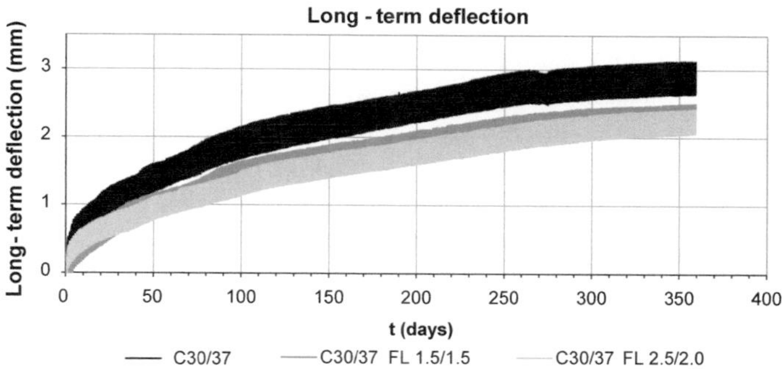

Fig. 8 Long term deflection for the beams from group 4

Table 3 Comparison between deflections of the beams from group 3 & 4 and creep ratios

n°	Concrete type deflection (mm)	C30/37	C30/37 FL 1.5/1.5	C30/37 FL 2.5/2.0
1	a_0 (inst.) g—group 3	0.845	0.83	0.815
2	a_0 (inst.) g—group 4	0.795	0.975	0.80
3	a (long-term)g—group 3	1.695	1.43	1.435
4	a (long-term)g—group 4	2.655	2.145	2.065
5	ratio: n° 4/n° 3	1.57	1.50	1.44
6	Creep ratio n° 3/n° 1	2.00	1.72	1.76
7	Creep ratio n° 4/n° 2	3.34	2.20	2.58

very important in order to find out the influence of the repeated variable load on the long term deflections. In the same table, the creep ratios for the beams from group 3 and 4 are presented separately. It is obvious that the addition of fibres reduced the creep effect.

3.2 *Time-Dependent Cracks Width*

Although all cracks that appeared during the application of the load or in the period of observing of 360 days, have been registered, the cracks width have been measured only in the middle third of the beams, i.e. in the part of the beams with constant bending moment.

As it was predicted, the cracks in all beams from group 3 (Fig. 9) appeared between the level of permanent and service load. At the reinforced concrete beams (C30/37), denoted as A_{31} and A_{32}, 4 and 6 cracks appeared from the precracking load, respectively. At the steel fibre reinforced concrete beams (C30/37 FL 1.5/1.5), denoted as B_{31} and B_{32}, 2 cracks appeared in each beam, while at the beams from C30/37 FL 2.5/2.0, denoted as C_{31} and C_{32}, 2 and 1 crack appeared. After removal of the variable load “q”, the cracks width decreased and increased thereafter up to the moment of appearance of new cracks. The mentioned new cracks were with smaller crack width and therefore they decrease the average value of all crack widths. This can be also noticed from the figures in the first 20 to 100 days. In total, at the beams A_{31} and A_{32}, 11 and 7 cracks developed, at beams B_{31} and B_{32}, 6 and 3 cracks developed, while at beams C_{31} and C_{32}, 4 and 8 cracks developed, respectively. After forming of all cracks there is only increase in their width.

From the Fig. 9, can be noticed that the instantaneous crack widths at the moment of precracking and after precracking for C30/37 FL 1.5/1.5 are bigger than C30/37, while for C30/37 FL 2.5/2.0 are significantly smaller. However, total crack width of C30/37 and C30/37 FL 1.5/1.5 are almost the same, while the one of C30/37 FL 2.5/2.0 is reduced for 31.8 %.

On the other hand, the long term crack widths of C30/37 and C30/37 FL 2.5/2.0 are similar, while the one of C30/37 FL 1.5/1.5 is decreased for 27.1 %. The long term crack widths of C30/37 FL 2.5/2.0 are of the same order of magnitude as C30/37 due to the fact that the instantaneous deflection of C30/37 FL 2.5/2.0 is much smaller and the cracks due to loading opened up later.

However, having in mind the complexity of the process of cracking, the randomly oriented fibres and the low level of stress from the permanent load, which is about 20 % from the concrete strength, can be concluded that steel fibres decrease the crack widths.

The beams from group 4 were first loaded with permanent load "g". At this load level there was no cracks appearance. Afterwards, the beams were loaded additionally with variable load "q". The variable load "q" was kept for 8 h and removed in the next 16 h. This was repeated every day in the period of 360 days. On certain time intervals, before and after application or removal of the variable load, measurements of the cracks width with crack microscope were performed.

In Fig. 10, time-dependent cracks width is presented as an average value of two beams. Dashed lines represent the cracks width at the level of permanent load "g", while the full lines represent the cracks width at the level of service load "g + q".

From Fig. 10, much clearer situation than for the group 3 can be noticed. This is due to the higher load level in the time intervals for 8 h each day for group 4.

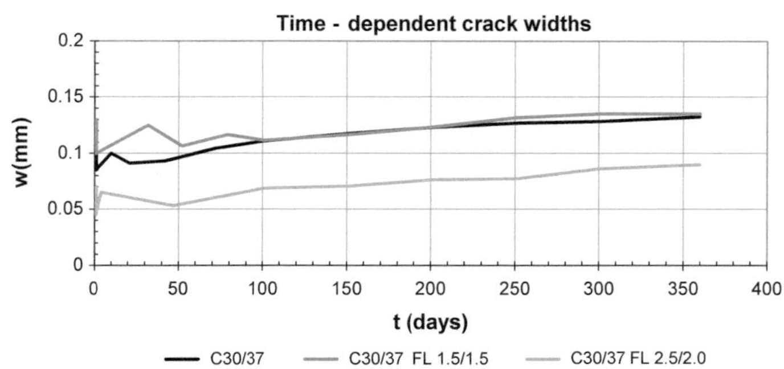

Fig. 9 Time-dependent cracks width for the beams from group 3

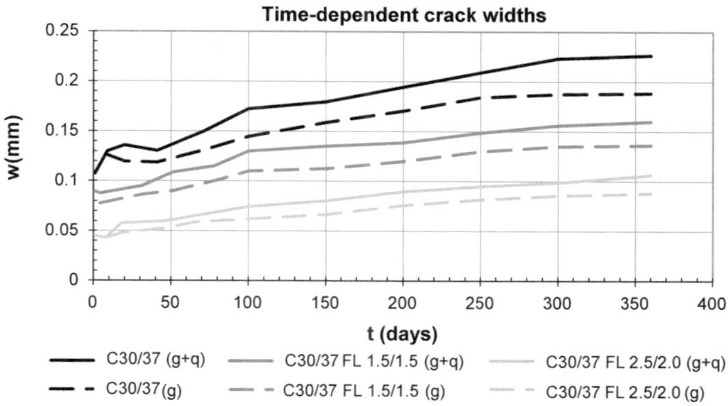

Fig. 10 Time-dependent cracks width for the beams from group 4

Instantaneous cracks width at the level of service load for concrete type C30/37 FL 1.5/1.5 are 17.6 % smaller than C30/37, while for C30/37 FL 2.5/2.0 are 58.4 % smaller than C30/37.

Total cracks width at the level of service load, when compared to C30/37, were decreased for 29.2 and 53.1 % for C30/37 FL 1.5/1.5 and C30/37 FL 2.5/2.0, respectively. At the level of permanent load similar decrease was obtained, 28.1 and 53.4 % for C30/37 FL 1.5/1.5 and C30/37 FL 2.5/2.0.

Long term cracks width at the level of service load, when compared to C30/37, were decreased for 40.3 and 47.9 % for C30/37 FL 1.5/1.5 and C30/37 FL 2.5/2.0, respectively. At the level of permanent load similar decrease was obtained, 42.0 % for C30/37 FL 1.5/1.5 and 46.9 % for C30/37 FL 2.5/2.0.

4 Conclusions

- The repeated variable load has significant influence on the time-dependent behaviour of reinforced and steel fibre reinforced concrete beams. The influence is slightly reduced with addition of fibres.
- From the presented results can be noticed that there is reduced ongoing of the fibre pull-out with time.
- Long term deflections of the beams subjected to sustained and repeated variable loads in the period of one year, for concrete types C30/37 FL 1.5/1.5 and C30/37 FL 2.5/2.0 are reduced around 20 % when compared to ordinary concrete C30/37, while the long term cracks width is reduced up to 40 and 50 % when compared to ordinary concrete C30/37.
- The results presented in this paper indicate the positive influence of additional reinforcing of RC beams with steel fibres regarding the creep effects.

Acknowledgments Many thanks to the Deutscher Academischer Austausch Dienst (DAAD), SEEFORM PhD studies and the Faculty of Civil Engineering-Skopje for the financial support.

References

1. Atanasovski S.: Influence of Long Term Actions to Limit States of Prestressed Concrete, Doctoral Dissertation. FCE-Skopje, Macedonia (1987)
2. Markovski G.: Influence of Variable Loads to Time-Dependent Behaviour of Prestressed Concrete Elements, Doctoral Dissertation. FCE-Skopje, Macedonia (2003)
3. Arangjelovski T.: Time-Dependant Behaviour of Reinforced High-Strength Concrete Elements Under Action of Variable Loads. FCE-Skopje, Macedonia (2011)
4. Tan K.H., Paramasivam P., Tan K.C.: Long term deflections of Steel fiber-reinforced concrete beams. ACI Struct. J. (1994)
5. Tan K.H., Saha M.K.: Ten-year study on Steel fiber-reinforced concrete beams under sustained loads, ACI Struct. J. (2005)
6. Vasanelli E., Micelli F., Aiello M.A., Plizzari G.: Long Term Behaviour of Fiber Reinforced Concrete Beams in Bending. BEFIB2012—Fibre reinforced concrete, Guimaraes (2012)
7. Nakov D.: Time-Dependent Behaviour of SFRC Elements Under Sustained and Repeated Variable Loads, Doctoral Dissertation. FCE-Skopje, Macedonia (2014)

Creep Behavior of a SFRC Under Service and Ultimate Bending Loads

D. Daviau-Desnoyers, J.-P. Charron, B. Massicotte, P. Rossi and J.-L. Tailhan

Abstract A research project was carried out to evaluate the flexural creep behaviour of SFRC beams under sustained service loads and the crack propagation under high sustained loads. The evolution of the deflection, the crack width and the crack propagation were measured in service conditions (CMOD w_s = 0.1 mm, load level P/P_s = 60 %) and ultimate conditions (w_u = 0.5 mm, P/P_u = 60–90 %). The results allowed assessing the impact of the initial CMOD and sustained load levels on creep, crack propagation, damage evolution, and the mechanisms leading to the rupture of the beams. In service conditions, the results show that the deflection and crack opening of SFRC beams under sustained loading stabilises quickly toward an asymptote and that the compliance remains constant. In ultimate conditions, the results show that crack propagation governs the failure mechanisms of SFRC beams subjected to high sustained load levels. Moreover, it was observed that the beams failure occurs when the state of damage defined by the static behaviour envelope is attained.

Keywords SFRC · Flexural creep · Crack propagation · Creep rate · Compliance · Static envelop

1 Introduction

Significant efforts have been made in the last decades to develop steel fibre reinforced concrete (SFRC) mixes, characterize their mechanical properties, standardize characterization tests, provide specific design rules and promote their structural

D. Daviau-Desnoyers
CIMA+, Montreal, Canada

J.-P. Charron (✉) · B. Massicotte
Polytechnique Montreal, Montreal, Canada
e-mail: jean-philippe.charron@polymtl.ca

P. Rossi · J.-L. Tailhan
Université Paris-Est, IFSTTAR, MAT, 75732 Paris, France

P. Serna et al. (eds.), *Creep Behaviour in Cracked Sections of Fibre Reinforced Concrete*, RILEM Bookseries 14,
DOI 10.1007/978-94-024-1001-3_18

applications [1–4]. Despite, further research regarding the long-term behavior of SFRC under a wide range of sustained load levels encountered in various conditions needs to be done to evaluate the influence of fibers on the amplitude and kinetics of creep as well as crack propagation.

Generally, macrocracks appear in structural elements throughout their service life. Yet, it is reasonable to assume that these macrocracks do not automatically compromise the serviceability of the structural element due to the ability of SFRC to withstand stresses upon the creation of a macrocrack. In this context, crack openings under sustained loading in service conditions should be limited to the range of SFRC efficiency. Moreover, crack openings must be maintained in ultimate conditions below a critical opening upon which the bridging effect of the fibres becomes negligible, or the loads must be kept under sustained load levels leading to uncontrolled crack propagation in SFRC [5, 6]. The propagation and the widening of macrocracks in a structural element under sustained loading must be predicted adequately in order to maintain the gain in durability provided by SFRC and avoid failure.

An experimental program on SFRC notched beams was carried out to provide insights on the mechanisms leading to the propagation of cracks and ultimately to the failure under sustained loading. The objectives were to assess creep, macrocrack propagation, and the evolution of the damage under service and ultimate loads. The experimental program included flexural creep tests on pre-cracked notched beams, both in sealed and drying moisture conditions at various crack openings and sustained loading levels. This paper focuses on the results obtained on beams in sealed and drying conditions submitted to an identical loading history. The loading history was selected to simulate service conditions and ultimate conditions found in a reinforced SFRC structure.

2 Experimental Program

2.1 Materials and Characterization

The mix used in the experimental program was a self-compacting SFRC, with a 70 MPa nominal compressive strength at 28 days. The mix contained Portland cement, sand, crushed aggregates with a continuous grading and a maximal diameter of 10 mm, high-range water-reducing admixture, water, and 78 kg/m^3 (1 %-volume) of Dramix ZP-305 steel fibres. These fibres are filaments of wire with a diameter of 0.55 mm, deformed into a hooked shape, and cut to a length of 30 mm giving an aspect ratio (l/d) equal to 55. The mixture proportions are summarized in Table 1.

Characterization specimens and flexural creep beams were unmolded after 24 h, then cured for 7 days at 100 % relative humidity, and finally were kept in air in a controlled environment room at 50 ± 5 % relative humidity and 23 ± 2 °C until

Table 1 Mixture proportions

Component	Quantity (kg/m^3)
Cement (CEM1 type)	650
Sand (calcareous)	843
Aggregate (calcareous)	622
Fibres (steel)	78.0
Superplasticizer	38.4
Water	195

they were tested. 7-day of humid curing is required in Canada for concrete structures exposed to freeze-thaw and chlorides [7]. Flexural creep beams were cured 6 months to minimize the variation of material properties due to concrete aging between the first and last flexural creep tests.

The mean mechanical properties at 28 days of the studied SFRC were: $f'_c = 75.7$, $f_t = 4.31$, $E_c = 35\ 800$ MPa, module of rupture (MOR) = 41.8 MPa. f'_c and E_c were measured on 3 specimens, f_t and MOR on 6 specimens.

2.2 *Flexural Creep Test Procedure*

Beams measuring 225 × 75 × 700 mm (dimensions selected according to the creep apparatus configuration) were successively submitted to a pre-cracking test to reach the specified crack mouth opening displacement (CMOD), then submitted to a creep test. The pre-cracking tests were conducted on a displacement controlled closed-loop testing system using the average beam mid-span deflection as the control signal; while the flexural creep tests were conducted in a modified compression creep rig with flat Freyssinet actuator loaded by a hydraulic accumulator (Fig. 1). The pre-cracking and the flexural creep devices had the same boundary conditions, 4-point bending configuration with 200 and 600 mm spans between the loading points and the lower supports respectively. Three roller supports and one pinned-roller support placed at one of the two loading points were used. In both tests, the deflection and the CMODs were measured respectively by an LVDT and a miniature spring return potentiometer on each side of the beam.

A typical loading procedure for a beam was selected to simulate service conditions and ultimate conditions found in a reinforced SFRC structure and is illustrated in Fig. 2 with the load-CMOD curve. The curve includes the pre-cracking phase (continuous line) to a specified CMOD before and after the maximum load P_{max} for part of the test in service conditions and in ultimate conditions respectively. The load measured at the specified CMOD in service, $w_s = 0.1$ mm, was called P_s, whereas the load measured at the specified CMOD at ultimate, $w_u = 0.5$ mm, was called P_u. The initial loading into the flexural creep frame (A), the sustained loading plateaus (B), the unloading-reloading cycles to estimate the compliance evolution during of sustained loading (C), and the final failure (★) are also presented on this figure. All creep tests were conducted until failure. The choice of $w_s = 0.1$ mm in

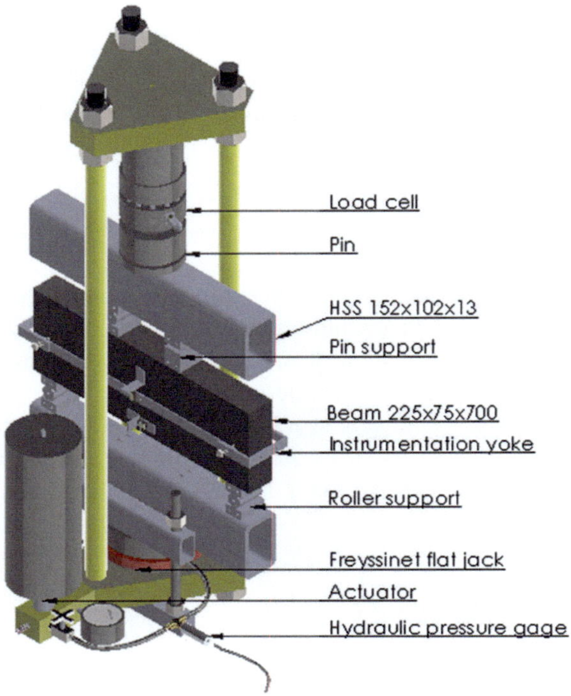

Fig. 1 Flexural creep test apparatus

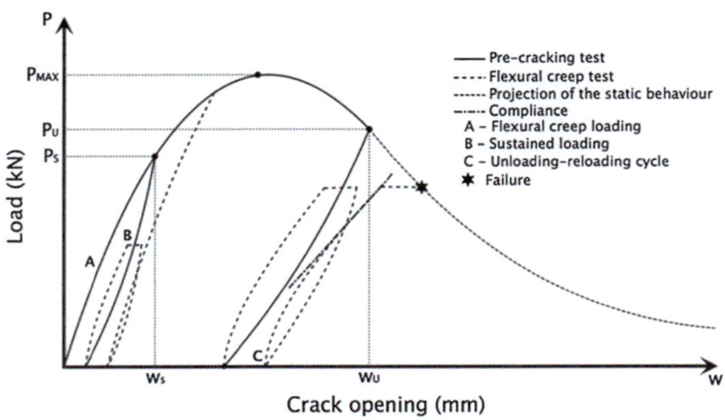

Fig. 2 Typical load-CMOD *curve* for high sustained loads

service conditions is related to the durability criterion in the Canadian Bridge Design Code [8], which limits crack opening to 0.25 mm. The choice of w_u = 0.5 mm in ultimate conditions is related to the fact that yielding of rebar occurs when crack width of 0.5–1 mm is achieved in reinforced SFRC structures [9].

Table 2 Loading characteristics

Service conditions	
CMOD (w_s)	0.1 mm
Loading level (P/P_s)	60 %
Loading period	4 weeks
Ultimate conditions	
CMOD (w_u)	0.5 mm
Loading level (P/P_u) and period	60 % (1 week), 75 % (2 weeks) and 90 % (1 week)

Table 2 summarizes the flexural creep test conditions such as the initial pre-cracked CMOD, the loading history and the loading period. Typically, the sustained load level (P/P_s) representing the service conditions was set to 60 % of P_s for 4 weeks, whereas the ultimate sustained load level (P/P_u) was increased from 60 to 90 % of P_u. Unloading-reloading cycle were performed every 7 days to evaluate the elastic compliance, which corresponds to the slope of the linear segment of the reloading part of the cycle. The compliance is an indicator of the damage state of the beam. The compliance measure should be understood as a measure of the length of a fictitious crack that would be mechanically equivalent to the real crack and its fracture process zone [10–12].

A total of ten beams were tested in the research project. Results of only 2 beams are presented in this paper (one in drying condition and one in sealed condition). Trends described for the 2 beams in the papers were also observed on all tested beams.

3 Results and Discussion

3.1 *Services Conditions*

3.1.1 Load-Deflection Responses

Figure 3a presents the load-deflection responses of the two beams, one in drying condition (P07-D) and one in sealed conditions (P07-S) cracked to a maximal crack opening w_s of 0.1 mm and then subjected to a sustained load level corresponding to 60 % of the load P_s during 4 weeks (Table 2). The results show that the load-deflection responses are comparable in terms of the initial stiffness and the residual deflection prior to sustained loading. The loads P_s of beams P07-D and P07-S measured at w_s are respectively 31.1 and 35.4 kN. Although the load P_s of beam P07-S is about 10 % greater than the load P_s of beam P07-D, the loads are also comparable as they are within the expected variability of the material.

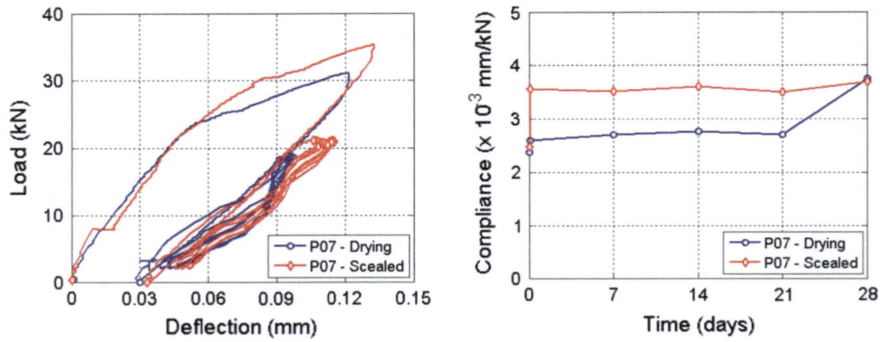

(a) Load-deflection responses (b) Compliance-time responses

Fig. 3 Mechanical behavior in service conditions

3.1.2 Deflection-Time and CMOD-Time Responses

Figure 4 presents respectively the deflection-time and the crack opening-time responses of the beams. The results show that the amplitude of the elastic deflection and the elastic crack opening of beam P07-S are slightly greater than beam P07-D. The elastic deflection and crack opening correspond to the deflection and crack opening due to the application of the sustained loading. This difference in amplitude is coherent to the fact that the load P_s of beam P07-S is greater than beam P07-D. Within the first 7 days of sustained loading, the deflection of beams P07-D and P07-S increases by approximately 4 and 7 % respectively, whereas the crack opening increases by approximately 9 and 17 % respectively. The deflection and crack opening, thus creep, of both beams stabilises toward an asymptote after 7 days of sustained loading.

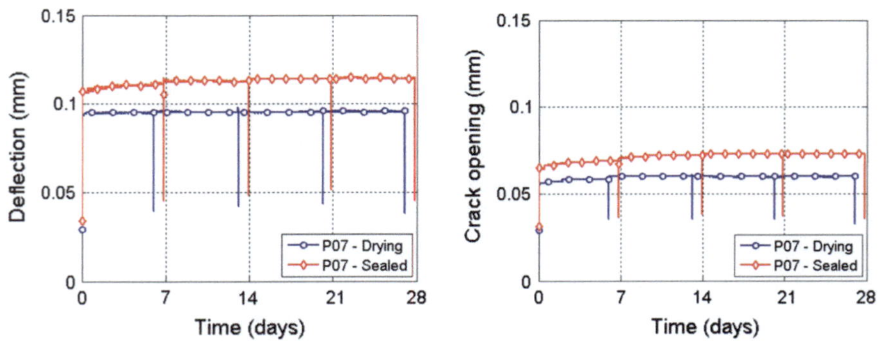

(a) Deflection-time responses (b) Crack opening-time responses

Fig. 4 Deformation in service conditions

3.1.3 Compliance Response

Figure 3b presents the compliance evolution of the beams. As expected, the compliance of beam P07-S at the time of application of the sustained loading is greater than the corresponding compliance of beam P07-D. This is due to the fact that the pre-cracking cycle of beam P07-S led to a greater deflection, thus a greater global damage state. Figure 3b also shows that the compliance evolution of both beams tends to be constant over the first 21 days. The compliance increase of beam P07-D observed between the 21st and 28th days of the test may be explained by the lack of precision in the compliance measurement. Indeed, the evolution of the deflection and CMOD, presented in Fig. 4, may not explain such a compliance increase within that time period.

3.1.4 Significance

The results show that the evolution of the creep behaviour of both beams, sealed (P07-S) and not sealed (P07-D), is comparable. Therefore, the influence of the moisture conditions on the evolution of the creep may be considered negligible for that condition.

Besides, the crack opening of the beams quickly stabilize to an average opening of 0.065 mm under sustained load representing 60 % of the load required to create opening of 0.1 mm. According to the Canadian Highway Bridge Design Code [8], the allowable crack opening is 0.25 mm in reinforced concrete. In addition, dead loads on bridge slabs associated to the sustained load represent around 10–30 % of the total service loads. Therefore, the results indicate that the use of SFRC in a bridge deck could be adequate to satisfy the crack opening limits defined by the code under sustained loading corresponding to its dead load. However, the impact of live loads (rapid cyclic loads) should be investigated.

3.2 Ultimate Conditions

After completing the 4 weeks of sustained loading in service conditions, the beams were loaded until the specified CMOD at ultimate conditions, $w_u = 0.5$ mm, was reached. The beams were then submitted to a sustained loading corresponding to 60 % of P_u for 7 days, after which the sustained loading was increase to 75 % for 14 days and finally to 90 % for 7 days (Table 2).

3.2.1 Load-Deflection Responses

The load-deflection curves of the two beams analysed in this paper are presented in Fig. 5. The loads P_u of beams P07-D and P07-S measured at w_u are respectively

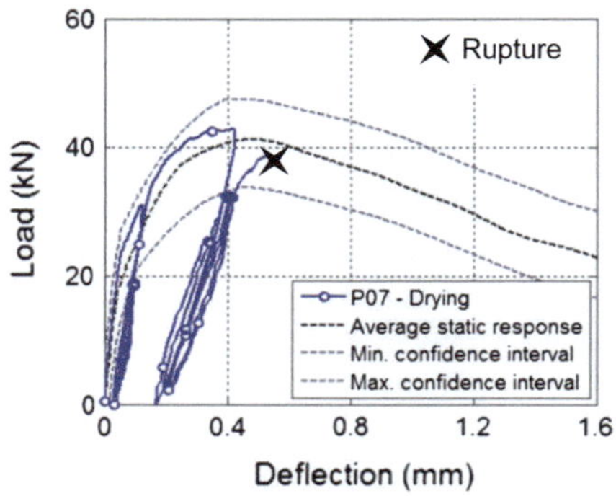

(a) Beam P07 – Drying

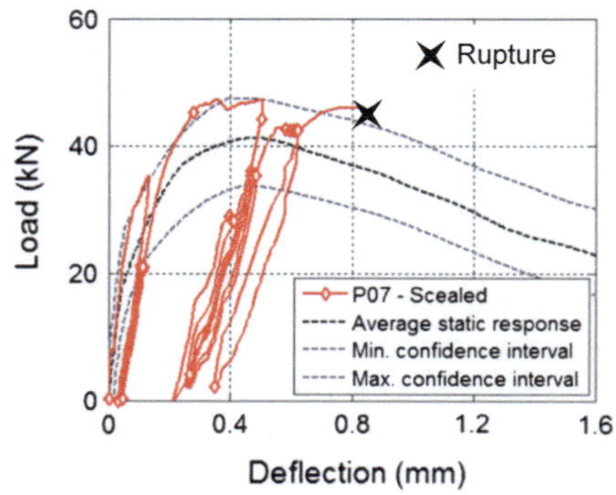

(b) Beam P07 – Scealed

Fig. 5 Load-deflection response at ultimate conditions

42.5 and 47.2 kN. The average load-deflection static bending response and its corresponding 95 % confidence intervals, calculated with the Student law and applied on 6 characterization tests, are also plotted in Fig. 5. The confidence intervals represent the static behaviour envelope of the characterization beams. In Fig. 5, a cross-mark indicates the failure point obtained during the ultimate loading conditions. The results show that the failure of beams occured as their deflection reached the static behaviour envelope defined by the 95 % confidence intervals. The complete experimental program revealed that this observation is true regardless of the CMOD (w_u) and the loading history (% P/P_u) applied on beam specimens [6, 13].

3.2.2 Deflection-Time and CMOD-Time Responses

The deflection-time and the CMOD-time responses of beams for the loading at ultimate conditions are presented in Fig. 6. The sustained load levels of each beam expressed in percentages of P_u are presented above the curves. The results show that beam P07-D and P07-S respectively failed upon the unloading-reloading cycle performed on the 49th and 56th day to reach 90 % of P/P_u. Although beam P07-D failed 7 days earlier than beam P07-S, Fig. 6 shows that the evolution of the deflection and the crack opening are comparable. As SFRC is an heterogeneous material, it is reasonable to assume that failure time of various specimens should vary under identical conditions. This phenomenon was also noted in studies by [5, 14].

Figure 6a, b were used to calculate the deflection and CMOD rates in the secondary creep phase for all sustained load intervals. The secondary rates were calculated for the time interval corresponding to the linear portion of each sustained load interval. In order to do so, mathematical functions best fitting the experimental results were defined for each sustained loading 7-day interval. Using these

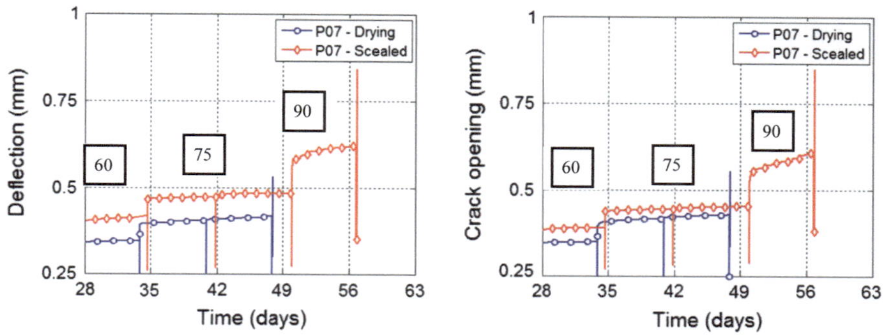

(a) Deflection-time responses (b) Crack opening-time responses

Fig. 6 Deformation at ultimate conditions

Fig. 7 Secondary rates per unit load versus load level *curves* [6]

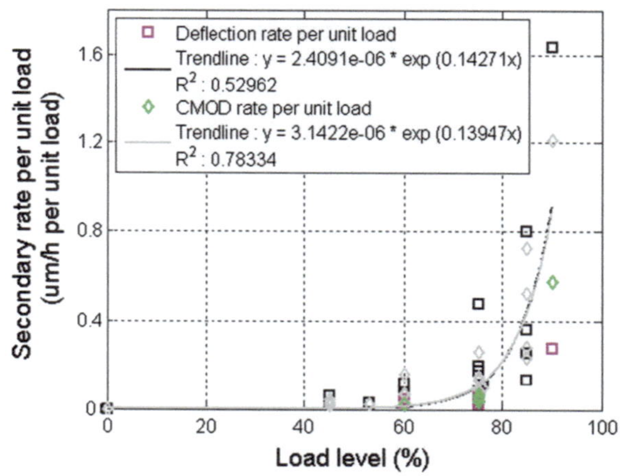

nonlinear functions, time intervals for which the second derivative approaches zero were selected. This method allowed determining accurately the boundaries of the linear portion of each sustained loading interval used to calculate the secondary creep deflection and CMOD rates.

The secondary deflection and secondary CMOD rates per unit load are plotted in Fig. 7 as a function of the sustained load level expressed in percentages of P_u. This figure includes the results obtained on five of the ten beams that were tested with the aforementioned procedure (Fig. 2) with various w_u and P/P_u ratios. A complete description of the results obtained for these five flexural creep tests may be found in [6, 13]. Figure 7 shows that an exponential relation exists between both the secondary creep deflection and secondary CMOD rates and sustained load levels greater or equal to 60 %. Yet, the deflection and the CMOD represent measurements at the global and at the local scale respectively. The deflection provides

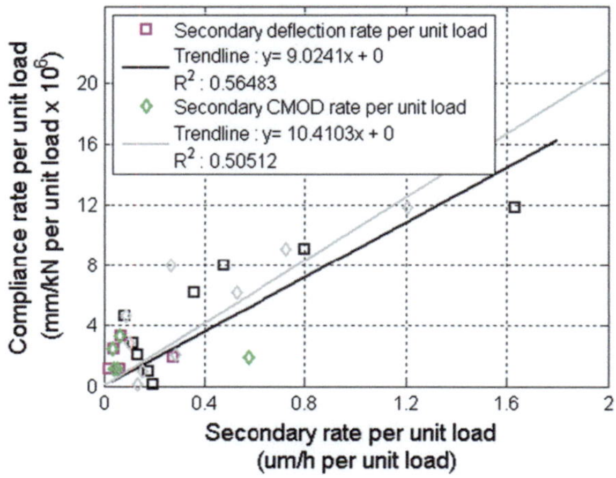

Fig. 8 Compliance rate versus secondary rates per unit load *curves* for high load levels [6]

global information on the beam's damage state, whereas the CMOD provides localised information of the macrocrack. Therefore, the increase in both the secondary creep deflection and secondary CMOD rates indicates that the same instigative mechanism of creep is involved at both the global and local scales.

The elastic compliance rate represents the difference of compliance between two unloading-reloading cycles divided by the elapsed time between those two cycles. The results are presented in Fig. 8 in function of the secondary deflection and secondary CMOD rates per unit load for sustained load levels greater than 60 % of the same series of specimens. This figure shows that a proportional relationship exists between the relative evolution of the compliance per unit load and the secondary creep deflection rate per unit load. The same remark applies to the secondary CMOD rates per unit load. In fracture mechanics, the elastic compliance corresponds to the overall damage state of the specimen. Hence, this correlation between the overall damage state and the secondary rates at high sustained load levels confirms that the propagation of the macrocrack governs the evolution of the long-term deflection of the SFRC beams. This conclusion was also made by Rossi et al. [15] in a similar study on ordinary concrete. Therefore, this propagation phenomenon must be directly linked to the cementitious matrix and not the fibres absent in ordinary concrete.

3.2.3 Significance

The observation on the beams' failure produced by high sustained load levels within the static behavior envelop goes along with the work of Walkinshaw [16] who suggested that a failure criteria based on a critical deflection could be applied for flexural creep tests performed on ordinary concrete. It means that a state of damage defined by the static behaviour envelope must be attained to cause the failure of the beams under high creep loads.

The previous correlations between deflection, crack opening and compliance under high sustained loads can be explained by the mechanism proposed by Rossi et al. [15, 17] for ordinary concrete. According to the authors, sustained loading leads to the creation and propagation of microcracks at the macrocrack tip. The creation of new microcracks generates local moisture shocks resulting in water and vapour pressure gradients. These gradients induce water movements from the capillaries surrounding the microcracks to the voids created by these microcracks, which results in additional tensile stresses due to this internal shrinkage. The additional tensile stresses lead to stress redistributions at the macrocrack tip that may lead to the creation of new microcracks or the propagation of the macrocrack, hence contributing to the widening of the macrocrack. This can be considered as the source of a significant portion of the long-term behaviour of creep.

In short, creep of concrete would be mainly due to microcracking, and the kinetics would depend on the water gradients. The physical explanation of this mechanism can be adapted to SFRC.

Test results, more precisely the proportional relation between the compliance and the secondary creep deflection and secondary CMOD rates, have shown the mechanism is directly linked to crack propagation. As fibres are present in the process zone and represent additional heterogeneity, it is reasonable to assume that their interfaces with the cement matrix are also affected by this phenomenon. New microcracks and the coalescence or propagation of microcracks in the vicinity of the fibres progressively deteriorate the fibre-matrix interface and may favor the macrocrack propagation, thus creep. However, creep of the fibre-matrix interface, acting at the local scale, does not play a significant role in the evolution of the compliance, which is mainly governed by the propagation of the macrocrack (structural scale). Therefore, the propagation of the macrocrack, favoured by the deterioration of the fibre-matrix interface, governs the secondary creep phase, ultimately leading to the beam's failure at high sustained load levels.

4 Conclusions

The results of the experimental campaign performed on pre-cracked SFRC beams submitted to sustained loading in service and ultimate conditions provided new understanding on the time-dependent deformation and the damage evolution.

The results of the flexural creep tests in typical service conditions (sustained loading of 60 % P/P_S, P_s obtained at $w_s = 0.1$ mm) have demonstrated that the evolution of the deflection, crack opening and compliance of beams remain very small to negligible. This should provide an extended durability to SFRC structures.

The results of the flexural creep tests in ultimate conditions (sustained loading of 60, 75 and 90 % P/P_u applied successively, P_u obtained at $w_u = 0.5$ mm) have shown that:

- a state of damage defined by the static behaviour envelope must be attained to cause the failure of the beams under creep loads over 45 % of their residual strength;
- an exponential relation exists between the secondary deflection and secondary CMOD rates per unit and sustained load level greater than 60 %, suggesting that the creep mechanism is the same regardless of the global or local scale of observation;
- a proportional relation exists between the evolution of the elastic compliance and the secondary deflection and CMOD rates for sustained load levels greater than 60 %.

Although the results on the studied SFRC are very promising, several types of SFRC are available in the industry. Therefore, further tests should be performed on several SFRC to confirm or to dispel the observed trends.

Acknowledgments This project has been financially supported by the Natural Sciences and Engineering Research Council (NSERC) of Canada, the Center for Research on Concrete Infrastructures of Quebec (FQRNT-CRIB). Materials were graciously provided by Bekaert, Holcim and Euclid. The authors gratefully acknowledge the technical staff of Polytechnique Montreal for its contribution in conducting the experimental program.

References

1. ACI Committee 544.: Design considerations for steel fiber reinforced concrete. In: ACI 544.4R-88, American Concrete Institute. ACI Farmington Hills (1996)
2. Barros J.: Fibre reinforced concrete: challenges and opportunities. In: Proceedings of the 8th International Symposium BEFIB. RILEM Publications, Bagneux, France (2012)
3. Reinhardt H.W., Naaman A.E.: High performance fibre reinforced cement composites (HPFRCC5). In: Proceedings of the International Conference of HPFRCC5. RILEM Publications, Bagneux, France (2007)
4. fib Bulletins 65–66.: Model Code 2010—Final Draft, vol. 1 and 2, pp. 720 (2012)
5. Zerbino, R.L., Barragán, B.E.: Long-term behavior of cracked steel fiber-reinforced concrete beams under sustained loading. ACI Mater. J. **109**, 215–224 (2012)
6. Daviau-Desnoyers, D., et Charron, J.-P., Massicotte, B., Rossi, P., Tailhan, J.-L.: Characterization of the propagation of a macrocrack under sustained loading in steel fibre reinforced concrete. Mat. Struct. **49**(3), 969–982 (2015)
7. CAN/CSA-A23.1-2.: Concrete: Constituents, Execution and Testing Methods. Canadian Standards Association, Mississaugo, Canada (2009)
8. CAN/CSA-S6-06.: Canadian Highway Bridge Design Code. Canadian Standards Association, Mississaugo, Canada (2013)
9. DeMontaignac, R., Massicotte, B., Charron, J.-P.: Design of SFRC structural elements: Flexural behavior prediction. Mat. Struct. **45**(4), 623–636
10. Rossi, P.: Fissuration du béton: du matériau à la structure—application de la mécanique linéaire de la rupture. In: PhD Thesis ENPC, pp. 240. Paris
11. Hu, X., Wittmann, F.: Experimental method to determine extension of fracture-process zone. J. Mater. Civ. Eng. **2**(1), 15–23 (1990)

12. Xu, S., Reinhardt, H.W.: Determination of double-K criterion for crack propagation in quasi-brittle fracture, part ii: analytical evaluating and practical measuring methods for three-point bending notched beams. Int. J. Fract. **98**, 151–177 (1999)
13. Desnoyers, D.: Characterization and modelization of crack propagation in fiber reinforced concretes under constant loading. In: PhD Thesis of Polytechnique Montreal, pp. 323. Montreal, Canada, (In french) (2015)
14. Barragán, B.E., Zerbino, R.L.: Creep behaviour of cracked steel fibre reinforced concrete beams. In: Proceedings of the 7th International Symposium, BEFIB. RILEM Publications, Bagneux, France (2008)
15. Rossi, P., Boulay, C., Tailhan, J.-L., Martin, E., Daviau-Desnoyers, D.: Macrocrack propagation in a concrete specimen under sustained loading: study of the physical mechanisms. Cem. Concr. Res. **63**, 98–104 (2014)
16. Walkinshaw, J.L.: Creep to Rupture Behavior of Concrete Beams. MSc Thesis of Massachusetts Institude of Technology, USA (1969)
17. Rossi, P., Tailhan, J.-L., Le Maou, F., Gaillet, F., Martin, E.: Basic creep behavior of concretes investigation of the physical mechanisms by using acoustic emission. Cem. Concr. Res. **42**, 63–71 (2012)

Durability of FRC Beams Exposed for Long-Term Under Sustained Service Loading

L. Candido, F. Micelli, E. Vasanelli, M.A. Aiello and G. Plizzari

Abstract Fiber Reinforced Concrete (FRC) is a material obtained by adding fibers to the concrete matrix. Fibers resist the crack opening and develop tensile residual strength. The extensive research activity carried out over the last decades on FRC has shown that such material has enhanced mechanical and durability properties as compared to plain concretes (PC) due to the cracking control. In fact, FRC flexural members show a higher number of cracks with reduced width; this occurrence involves an improved mechanical behaviour mostly under service conditions and a higher durability due to the reduced attacks of aggressive agents and to the more controlled effect of long terms phenomena. The experimental study presented herein discusses the long-term behaviour of FRC flexural beams subjected to sustained service load and environmental exposure up to 72 months. The effects of different short fibers (polyester and steel), sustained loading and aging were investigated. The results show the beneficial effects of fibers in terms of reduced crack width and increased flexural stiffness. The experimental data on cracking behaviour are finally compared with the analytical predictions obtained according to the formulation provided by fib Model Code 2010.

Keywords Concrete · Concrete · Concrete · Concrete

L. Candido · F. Micelli · M.A. Aiello
Department Engineering for Innovation, University of Salento, Salento, Italy

E. Vasanelli
CNR-IBAM, Lecce, Italy

G. Plizzari (✉)
DICATAM, University of Brescia, Via Branze 43, 25123 Brescia, Italy
e-mail: giovanni.plizzari@unibs.it

P. Serna et al. (eds.), *Creep Behaviour in Cracked Sections of Fibre Reinforced Concrete*, RILEM Bookseries 14,
DOI 10.1007/978-94-024-1001-3_19

1 Introduction

The presence of short fibers in the concrete mix helps controlling or delaying cracking [1, 2] and developing moderate time-dependent effects [3–5]. FRC members show improved toughness and ductility under dynamic and seismic loading, as well as under static and fatigue loading. Moreover, FRC members show a higher number of cracks with a reduced width, as compared to plain concrete elements. Therefore, the material may develop an enhanced protection against water and contaminants, which are responsible for corrosion of rebars and potential deterioration of concrete. Such behaviour offers the perspective of a longer service life, especially in aggressive environments.

There are still limited analytical models concerning crack width and spacing in FRC structural elements under sustained bending loads due to a lack of experimental data [4, 6]. A crack width relationship for FRC elements is desirable for designers and engineers all over the world. RILEM TC 162 TDF [7] and fib Model Code (MC) 2010 [3] have proposed equations considering this issue. However, those models need further validation by means of experimental results. Furthermore, the effects of long-term loading on crack width in FRC elements is a crucial topic nowadays.

This research addresses the cracking behaviour of FRC beams reinforced with traditional rebars and stirrups. Results concern an experimental campaign on full-scale FRC beams, tested under long-term bending loading. The applied sustained load corresponds to about 50 % of the ultimate load, in order to reproduce service conditions. After months of exposure, the FRC beams have been tested under a four-point bending configuration for determining the residual stiffness as well as the cracking behaviour. During the test, displacements and crack widths were measured at different loading steps.

2 Experimental Campaign

The work aimed at understanding the effects of long-term loads on the performance of RC elements with fibers. The experimental campaign reported in this paper follows the research study published in [8]. Thirteen full scale beams were cast and left in laboratory for one year. The first group, including three reinforced concrete beams cast respectively with plain concrete, steel fiber reinforced concrete and polyester fiber reinforced concrete, was tested in laboratory (_L). These beams were used as reference specimens. The remaining beams were exposed to natural weathering (_E) and subjected to sustained loads in order to simulate real service conditions. They can be distinguished into two sets of five beams that were respectively put in two loading frames: the first was kept under sustained loads for 17 months (_17), the last for about 72 months (_72). Both sets consisted of one beam casted with plain concrete (_PC), two beams casted with steel fiber reinforced

Table 1 Beam ID tags, time of exposure and testing

Beam ID tag	Time of exposure [months]	Testing time from casting [months]
PC_L/ST_L/POL_L	–	12
PC_E_ST/ST_E_ST/POL_E_ST	17	29
PC_E_LT/ST_E_LT/POL_E_LT	72	84

Table 2 Experimental mechanical properties of steel rebars

Reinforcing element	Diameter (mm)	Yield strength (MPa)	Ultimate strength (MPa)	Elongation at rupture (%)
Longitudinal bar	14	520	614	12.2
Stirrups	8	567	600	4.8

Table 3 Geometrical and mechanical characteristics of the short fibers

Type	Shape	Diameter [μm]	Length [mm]	L/D	Tensile strength [MPa]	Elastic modulus [kN/mm^2]
Steel	Hooked	600	30	50	>1150	210
Polyester	Waved	450	30	66	400–800	11.3

concrete (_ST) and two beams casted with polyester fiber reinforced concrete (_POL). Table 1 reports the details of these beams such as ID tags, time of exposition, and testing time from casting.

The investigation of the long-term effects concerned two steps besides the material characterization. A first step consisted in field-tests aimed at detecting the cracking evolution (namely position, width and length of cracks) of the exposed beams under sustained loads. A second laboratory phase concerned four-point bending tests on beams up to their failure.

Table 2 reports the average mechanical properties of longitudinal bars and stirrups employed as beam reinforcement; they were measured according to UNI EN ISO 15630-1 [9] by testing three samples for each diameter.

Two different types of fibers were used: polyester fibers (PE30) and steel fibers (60 × 30). The aspect ratios (length/diameter) were equal to 50 and 66 for steel and polyester fibers, respectively. All the geometrical and mechanical properties of the short fibers, as measured by the manufacturer, are summarized in Table 3. As a consequence, three different concrete mixes were prepared: a reference mix without fibers (PC), a concrete mix with a volume fraction of 0.6 % of steel fibers (ST) and a concrete mix with 0.9 % of polyester fibers (POL). The choice on fiber contents imply equivalent total weight. The fiber volume fractions were set according to the results obtained by a previous research [10], considering the best compromise in terms of mechanical properties and workability of concrete. All the mixes had a

Table 4 Concrete mix

	ST	POL	PC
CEM 32.5R II-A/LL [kg/m^3]	300	300	300
Superplasticizer CRTV-L [kg/m^3]	1.59	2.5	1.77
Sand (0–4) [kg/m^3]	1028	1023	1037
Gravel (4–10) [kg/m^3]	704	701	710
Water/cement ratio	0.65	0.65	0.65
Fiber content	0.6 %	0.9 %	–

Table 5 Cubic cylindrical compressive strength

Beam ID tag	Average cubic strength [MPa]	COV (%)	f_{R1} [MPa]
PC	25.80	4	–
ST	21.40	8	0.88
POL	23.20	8	0.40

water-cement ratio equal to 0.65, a cement type 32.5R II-A/LL (UNI EN 196-1 [11], see Table 4) and a workability class S5. Four cubes (150 × 150 × 150 mm) for each mix (PC, ST and POL) were cast for quality control; the values of the compressive strength obtained after 28 days from casting are reported in Table 5. The compressive strength of ST mix was comparable to that of POL mix and both were slightly lower than that evaluated for PC. This was probably due to a higher content of entrapped air in the mix as it was found in a previous research carried out by the Authors [10]. The influence of fibers on compressive strength can be considered negligible, especially for low fiber contents (ACI Committee 544 [12]). Table 5 reports indeed the value of f_{R1}, as defined in Model Code 2010 [3]. These values are determined by means of inverse analysis adopted to fit the load-crack mouth opening curve of notched specimens subjected to a four-point bending test.

The two loading frames at the exposure site are depicted in Fig. 1. The loading scheme represents a four point bending condition with a span of 280 cm and a distance of 90 cm between the two loading points.

The reinforced concrete beams were designed according to the Eurocode 2 [13], on the basis of the loading scheme reported in Fig. 2a. The beams were 3000 mm long, with square cross section 250 × 250 mm. Three and two ϕ14 mm bars were respectively used as longitudinal reinforcement in tension and compression sides of the beams. In the transverse direction, 8 mm diameter stirrups were spaced 140 mm in the central part of the member and 70 mm close to the supports. The reinforcement ratio was chosen to have a bending failure of the beam, with concrete crushed after steel yielding of the tension bars. Vertical stirrups were designed to prevent premature shear failure. Figure 2a shows the geometry of the beams as well as the reinforcement details, while Fig. 2b show the actual experimental set-up.

The two loading frames were located in Brindisi (Italy), in an industrial area that is 600 m far from the sea. The site has a high relative humidity rate (>70 %) and

Fig. 1 The two exposure loading frames

presence of chlorides in the air. According to EN 206-1 [14], the exposition class of the site is XC3 for corrosion induced by carbonation, and XS1 for corrosion induced by chlorides. Five beams were loaded in each frame by means of a screw jack. A service load of about 50 kN has been applied during the exposure time, accounted as about 50 % of the ultimate design load. Due to relaxation effects of the loading system, the beams were periodically reloaded in order to keep the sustained load constant. During the exposure time, a handheld digital microscope with 200× magnification was used to measure crack width at the bottom edge of the beams.

After one year from casting, the laboratory beams (control specimens) were tested according to the four point scheme represented in Fig. 2. During the test, deflections at midspan and at quarter points of the beams and crack widths were continuously monitored respectively by means of three Linear Variable Differential Transducers (LVDTs) and a digital microscope. Tests were performed under load control; a load cell having a capacity of 30 tons was used to measure the applied force. The crack pattern of beams (number of cracks, crack widths and crack lengths) was recorded at five load steps: Step 1 = 20; Step 2 = 30; Step 3 = 50; Step 4 = 80; Step 5 = 100 kN. Beams under sustained load were tested up to failure after 29 and 84 months from casting, respectively, according to the same protocol adopted for laboratory beams.

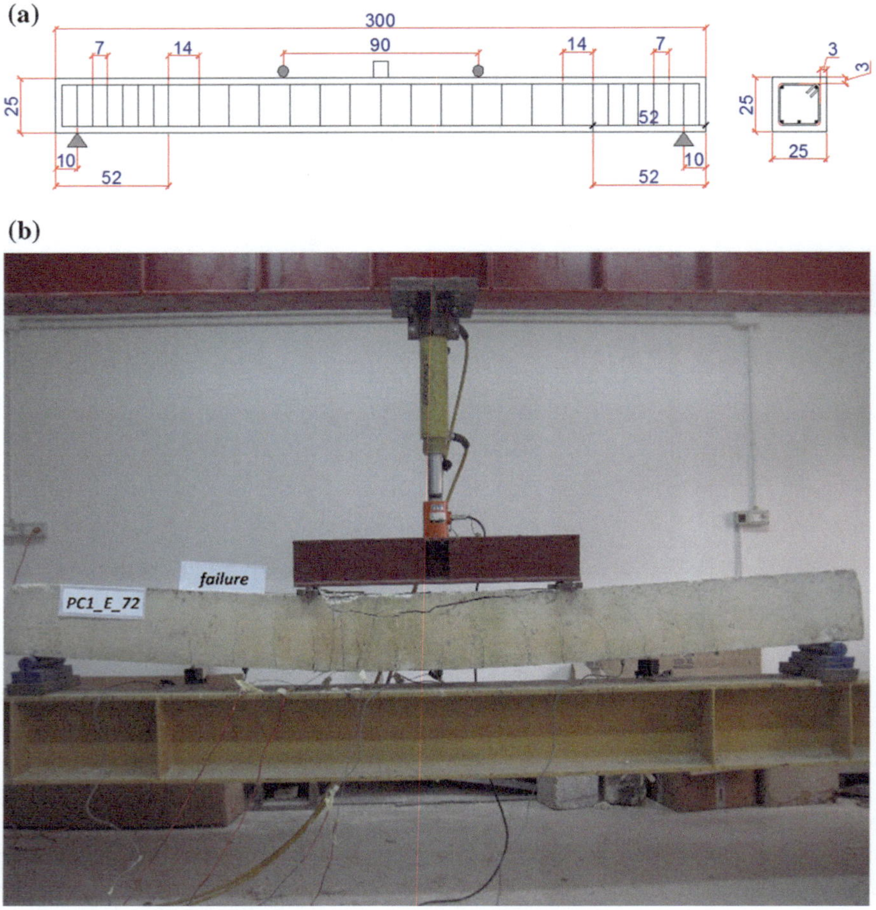

Fig. 2 **a** Beam details (cm) and **b** experimental set-up

3 Results and Discussion

Experimental results of the tests carried out on beams exposed for 17 months have been already presented in Vasanelli et al. [8] while a preliminary analysis of the mechanical behavior of the members exposed for 72 months is presented in [15].

During the four-point testing phase, as expected, all the thirteen beams failed by concrete crushing after yielding of tensioned steel bars, just as designed by the Authors; PC beams presented the most extensive concrete damage at the compressed side, with concrete spalling. Figure 3 shows the load—midspan deflection curves of the tested beams grouped by mix type and time of exposure. Figure 3a, b refer to steel and polyester FRC beams, respectively. In the same figures, the load-deflection curve related to the reference PC beams (in black) are also reported.

The figure shows how, after 6 years (72 months) of exposure and sustained loading, FRC beams still perform better (as is the case of steel FRC) or similarly (as is the case of polyester FRC) than the unexposed plain concrete beam. Therefore, the beneficial effect of fibers in the long-term behavior of conventionally reinforced members exposed under sustained loading can be interpreted as a delayed start of the time-dependent effects with respect to the corresponding plain concrete members. It shall be noticed, at the same time, that the influence of fibers in terms of bearing capacity is almost negligible due to the limited size of the beam and the presence of a normal amount of traditional longitudinal reinforcement. Indeed, it can be observed that, after 72 months of sustained loading, there is a limited difference in the global performance of the PC and FRC beams.

Table 6 reports the results of bending tests on beams in terms of secant stiffness, computed between null load and 50 kN, midspan deflections at each loading step, including failure, ductility value and ultimate and steel yielding loads. The ductility value has been evaluated as the ratio between the deflections at failure and at steel yielding. The flexural stiffness of all 72-month exposed FRC beams is greater than that of the reference PC beams, especially for SFRC beams where the increment is as high as 12–21 %, while polyester FRC beams show lower increments (3–17 %). For 17-month exposed FRC beams the stiffness values are even higher, as already evidenced in [8]. By the way, the stiffness values of all 72-month exposed beams are close to the corresponding (unexposed) laboratory beams. Indeed, exposed beams show a lower value of stiffness up to the load level applied during the exposure period. Since such effect of applied loading is most pronounced for the PC beam, as it can be deduced by Fig. 3, then the presence of fibers is beneficial to contain the effects of sustained loading as well. Above a load of about 80 kN no time-dependent effect is evidenced for all beams. The contribution of fibers in the elastic behavior of beams is also proved by the general increment in the value of loading corresponding to the steel yielding, which is about 2.1 and 5.5 % for POL

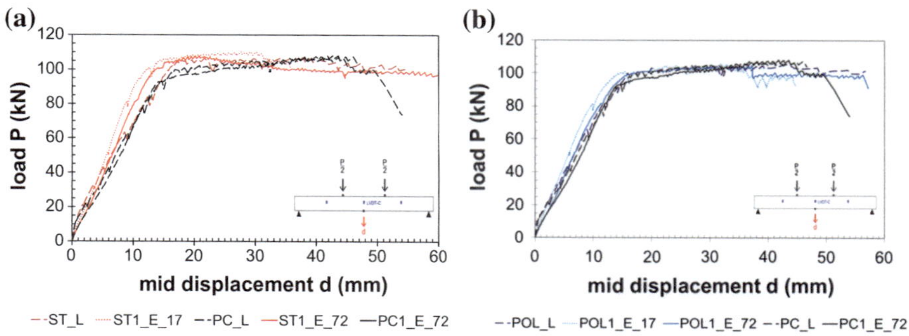

Fig. 3 Load versus midspan deflection *curves* comparisons among FRC beams and the reference PC beam. **a** Steel FRC. **b** Polyester FRC

Table 6 Mechanical behavior of beams in the four-point bending test

Beam ID tag	Stiffness [kN/m]		Midspan deflection [mm]					Ductility	Ult. load (kN)	Steel Yield. load (kN)
	0–50 kN	Δ[a]	20 kN	30 kN	50 kN	80 kN	Failure			
PC_L	7027	–	1.94	3.99	7.1	12.39	45.42	3.51	107.62	84.24
POL_L	7246	3 %	1.42	2.74	6.90	11.55	41.38	3.14	105.52	88.03
ST_L	7849	12 %	1.54	3.06	6.37	11.67	53.36	3.78	107.12	91.43
PC1_E_17	6303	–	3.53	5.08	7.93	11.47	46.1	3.47	104.42	88.33
POL1_E_17	8865	41 %	2.03	3.35	5.64	9.55	44.61	4.01	105.82	87.84
POL2_E_17	9016	43 %	2.20	3.36	5.54	8.88	37.92	3.68	110.92	90.23
ST1_E_17	8956	42 %	2.27	3.40	5.58	8.63	35.82	3.46	110.42	91.93
ST2_E_17	9891	57 %	1.56	2.80	5.05	8.26	38.16	3.93	117.41	92.93
PC_E_72	6312	–	3.03	4.80	7.92	11.86	49.5	3.54	108.12	92.50
POL1_E_72	7365	17 %	2.79	4.16	6.78	10.97	56.47	4.03	104.72	94.93
POL2_E_72	6554	4 %	3.20	4.69	7.62	12.00	49.75	3.25	104.42	94.13
ST1_E_72	7653	21 %	2.72	4.04	6.53	10.13	65.7	5.39	107.92	95.23
ST2_E_72	7210	14 %	2.98	4.37	6.93	10.8	46.99	4.33	110.42	98.73

[a]Δ = percentage variation of stiffness with respect to PC beams

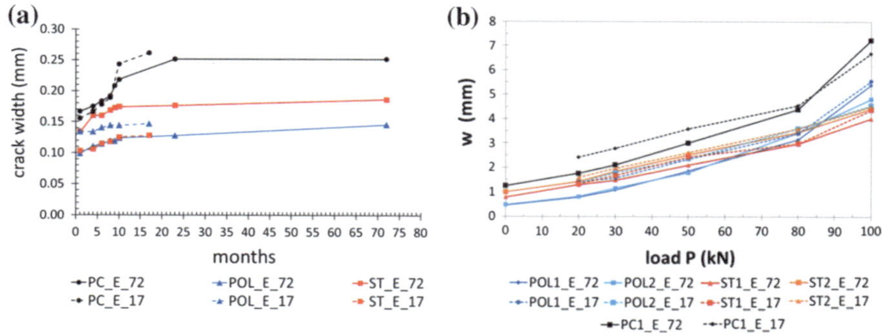

Fig. 4 Mean crack width at the constant moment zone of exposed beams: (**a**) measured in situ (**b**) measured in laboratory

and ST beams respectively, compared to the PC beam belonging to the same group. By the way, the contribution of fibers to the ultimate load is not as important in conventionally reinforced concrete beams, according to available literature [16–18]. Furthermore, the values of ductility for FRC beams show a very limited increment for polyester beams while SFRC beams reach an increment as high as 20 %. In fact, computed as average by beam types on all groups (Table 6), the mean values of ductility are 3.5, 3.62 and 4.18 respectively for PC, POL and ST beams.

Figure 4 reports the mean crack width along the constant moment zone of exposed beams. Figure 4a specifically shows the mean crack width measured on-site during the 72 months of exposure under sustained loading, while Fig. 4b shows the values of the mean crack width registered during the four-point bending test carried out in the laboratory. The values of crack widths corresponding to a null load (0 kN) are the residual ones, referring to the exposed beams after unloading. Under sustained loading, the crack openings stabilize after 25 months for PC beams and much earlier, after about 10 months, for FRC beams. The use of fibers in concrete members helps in reducing the crack openings, as much as 50 % with an amount of 0.9 % of polyester fiber, in both loaded or unloaded conditions. However, as the load intensity increases, steel fibers prove to be more effective in reducing crack openings, although they are used in a lower volume fraction (0.6 %), as compared to polyester fibers (0.9 %).

Table 7 reports the number of cracks, the mean crack spacing and the mean crack width for all the exposed beams, referring to a load intensity level of 50 kN and to the constant moment zone. The number of cracks occurred in the PC beams is always 7, corresponding to the number of stirrups present in the constant moment zone. FRC beams have an equal or lightly higher number of cracks along the same length (7–9), including possible crack splitting. Furthermore, in FRC specimens the total crack width—obtained as number of cracks times mean crack width—accounts for a reduction up to 30 %, as compared to that of the corresponding PC beams.

Table 7 Number of cracks, mean crack spacing and total crack length in the constant moment zone

Beam ID tag	Total # cracks	Mean cracks spacing [cm]	Δ_1^a	Mean crack width [mm]	Beam ID tag	Total # cracks	Mean cracks spacing [cm]	Δ_2^b	Mean crack width [mm]
POL1_E_72	8	11.69	−16.32 %	0.138	POL1_E_17	7	14.18	−0.14 %	0.189
POL2_E_72	7	14.19	1.57 %	0.154	POL2_E_17	8	12.21	−14.01 %	0.167
ST1_E_72	7	14.12	1.07 %	0.181	ST1_E_17	8	10.23	−27.96 %	0.178
ST2_E_72	7	14.09	0.86 %	0.195	ST2_E_17	9	10.50	−26.06 %	0.169
PC1_E_72	7	13.97	–	0.239	PC1_E_17	7	14.20	–	0.302

[a] Δ_1 = difference between the values of the crack spacing respectively for the beam under examination and the reference PC1_E_72 beam

[b] Δ_2 = difference between the values of the crack spacing respectively for the beam under examination and the reference PC1_E_17 beam

Table 8 Analytical predictions (MC2010) in bold for crack spacing and width; experimental data in brackets

Beam ID tag	Short-term		Long-term, stabilized cracking stage	
	Spacing S_{rm} [mm]	Width w_d [mm]	Spacing S_{rm} [mm]	Width w_d [mm]
PC_E_72	128 (141)	0.235 (0.160)	128 (141)	0.329 (0.252)
POL_E_72	107 (131)	0.182 (0.117)	107 (131)	0.262 (0.142)
ST_E_72	82 (123)	0.132 (0.119)	82 (123)	0.196 (0.175)

Table 8 reports in bold the analytical predictions of the values of mean crack spacing and width, according to Model Code 2010 [3]; the experimental data are reported in brackets. As it can be noticed, the predictions on RC specimens can be considered as the most reliable, while for FRC specimens the predictions underestimate quite much the beneficial effect of fibers in reducing crack width. Furthermore, the presented experimental campaign did not show time-dependency in the number of cracks, and therefore in the mean crack spacing, at the constant level of applied load. According to the experience of the Authors, the number of cracks only varies with the level of the applied load. Furthermore, in all the tested beams the crack spacing is strongly influenced by the stirrup spacing, as most cracks have formed in correspondence of the effective position of stirrups within the member. The analytical prediction is not able to catch this phenomenon.

4 Conclusions

The experimental study presented herein aimed at understanding the effects of long-term loading on the performance of conventionally reinforced RC/FRC beams exposed to natural weathering under sustained loadings, at its maximum serviceability level of intensity. Afterwards, the beams were unloaded and tested in the laboratory up to failure. The crack pattern and displacements were accurately detected during both long-term loading and final laboratory tests until failure.

The following observations can be drawn. The flexural stiffness of all FRC beams is slightly greater than that of the reference PC beams, whereas steel FRC beams show the best mechanical performances in terms of bearing capacity and stiffness. The most pronounced effect of stiffness loss due to the sustained loading occurs for the PC beam, while the presence of fibers seems enough beneficial to contain such effects. Indeed, the presence of fibers leads to an improved ductility in structural members (especially in the case of for steel FRC) and a greater steel yielding load, as well as smaller midspan deflections.

The crack opening develops over time, thus requiring the introduction in the building codes of the time for measuring the crack opening in order to avoid controversies. In fact, under sustained loading, the crack openings stabilize after 25 months for PC beams and much earlier, after about 10 months, for FRC beams.

The use of fibers in concrete members helps in reducing the crack openings, as much as 50 % with an amount of 0.9 % of polyester fiber, in both loaded or unloaded conditions. Polyester FRC beams show lower values of mean crack width, as compared to steel FRC beams, for loading level up to about 60 kN (above the serviceability intensity level). However, as the load intensity increases, steel fibers prove to be more effective in reducing crack openings, although they are used in a lower volume fraction (0.6 %), as compared to polyester fibers (0.9 %). Polyester and steel fibers have different shapes, respectively prismatic and circular, and different mechanical properties besides different volume contents; all these parameters should be properly taken into account, in order to assess reliable relationships useful from a design point of view.

Finally, the analytical predictions obtained by means of fib Model Code 2010 instructions are reliable for RC members only, while for FRC members the beneficial effects of fibers are underestimated. The strong influence of stirrups in the cracking behavior of RC beams should be addressed as well.

In summary, FRC beams prove to have an improved mechanical and cracking behavior than plain concrete ones as fibers appears also effective in limiting time-dependent phenomena. Therefore, the beneficial role of fibers may be considered relevant in terms of structural durability, since the reduction of crack width, even in the long-term conditions, involves an increase of the structure service life. A deeper analysis of results, in future research developments, may allow assessing valuable design indications.

Acknowledgments The Authors are grateful to Mr. Giuseppe de Rossi and to La Matassina Technology for providing both polyester and steel fibers.

References

1. Di Prisco, M., Plizzari, G., Vandewalle, L.: Fibre reinforced concrete: new design perspectives. Mater. Struct. **42**, 1261–1281 (2009)
2. Norme tecniche per le costruzioni, D.M.: 14 Jan 2008 (2008)
3. FIB Model Code.: First Complete Draft, vol 2, Chaps. (7–10)' in fib Bulletin 562010, ISBN 978-2-88394-096-3 (2010)
4. Tan, K.H., Paramasivam, P., Tan, K.C.: Cracking characteristics of reinforced steel fiber concrete beams under short and long-term loadings. Adv. Cem. Based Mat. **2**, 127–137 (1995)
5. Altoubat, S.A., Lange, D.A.: A new look at tensile creep of fiber reinforced concrete. ACI Spec. Publ. **216**, 143–160 (2003)
6. Vasanelli, E., Micelli, F., Aiello, M.A., Plizzari, G.: Crack width prediction of FRC beams in short and long term bending condition. Mat. Struct. **47**(1–2), 39–54 (2014)
7. Vandewalle, L.: Rilem technical committees—recommendations of RILEM TC 162-TDF: Test and design methods for steel fibre reinforced concrete. Mat. Struct. **33**(225), 3–5 (2000)
8. Vasanelli, E., Micelli, F., Aiello, M.A., Plizzari, G.: Long term behaviour of FRC flexural beams under sustained load. Eng. Struct. **56**, 1858–1867 (2013)
9. UNI EN ISO 15630-1. Steel for the reinforcement and pre-stressing of concrete—test methods —part 1: reinforcing bars, wire rod and wire. Italian Board of Standardization (UNI)

10. Vasanelli, E., Micelli, F., Aiello M.A. and Plizzari, G.: Mechanical characterization of fiber reinforced concrete with steel and polyester fiber. In Proceedings of the BEFIB 2008, 7th RILEM International of Symposium on Fibre Reinforced Concrete, pp. 537–546. Chennai, India (2008)
11. UNI EN 196-1.: Cement—Part 1: Composition, Specifications and Conformity Criteria for Common Cements, Italian Board of Standardization (UNI) (2007)
12. ACI Committee 544.: Measurement of Properties of Fiber Reinforced Concrete. ACI Report 544.2R-89. American Concrete Institute, Farmington Hillis
13. Eurocode 2.: Design of Concrete Structures—Part 1–1: General Rules and Rules for Buildings. EN 1992-1-1 (2004)
14. UNI EN 206-1.: Concrete—Specification, Performance, Production and Conformity, Italian Board of Standardization (UNI), 2001 (2010)
15. Candido, L., Micelli, F., Vasanelli, E., Aiello, M.A., Plizzari, G.: Cracking behaviour of FRC beams under long-term loading. Concreep **10**, 1147–1156 (2015)
16. Swamy, R.N., Sa'ad, A.: Deformation and ultimate strength in flexure of reinforced concrete beams made with steel fiber concrete. J. Proc. **78**(5) (1981)
17. Tan, K.H., Paramasivam, P., Tan, K.C.: Instantaneous and long-term deflections of steel fiber reinforced concrete beams. ACI Struct. J. **91**(4), 384–393 (1994)
18. Abdul-Ahad, Ramzi B., Aziz, O.Q.: Flexural strength of reinforced concrete T-beams with steel fibers. Cem. Concr. Compos. **21**(4), 263–268 (1999)

MIX
Papier aus verantwortungsvollen Quellen
Paper from responsible sources
FSC® C105338

If you have any concerns about our products,
you can contact us on
ProductSafety@springernature.com

In case Publisher is established outside the EU,
the EU authorized representative is:
Springer Nature Customer Service Center GmbH
Europaplatz 3, 69115 Heidelberg, Germany

Printed by Libri Plureos GmbH
in Hamburg, Germany